I0192315

The Way of the Air

The Way of the Air
Aircraft & Airmen of the First World War
1914-1918

Edgar C. Middleton

LEONAUR

The Way of the Air
Aircraft & Airmen of the First World War 1914-1918
by Edgar C. Middleton

First published under the title
The Way of the Air

Leonaur is an imprint of Oakpast Ltd

Copyright in this form © 2012 Oakpast Ltd

ISBN: 978-1-78282-022-2 (hardcover)
ISBN: 978-1-78282-023-9 (softcover)

http://www.leonaur.com

Publisher's Notes

The views expressed in this book are not necessarily
those of the publisher.

Contents

PART 3: OTHER CRAFT AND THE FUTURE

DEDICATION

To the Memory of Friends Who Have Fallen
In the Great Fight

Captain Adrian Liddell, V.C., R.F.C.

Flight Sub-Lieut. R. A. J. Warneford, V.C., R.N.

Flight Lieut. Rosher, R.N.

Flight Lieut. Talbot, R.N.

Flight Lieut. Graham, R.N.

Flight Commander Beard, R.N.

Captain Basil Hallam Radford, R.F.C.

And

Second-Lieut. Arthur Fisher, R.F.C.

"Who Found Glory Only Because
Glory Lay In the Plain Path of Duty"
This Book is Dedicated

Author's Note

The idea of this little book is to give as clear and graphic a description of modern aviation as circumstances will permit; of the new heroic race of men to which flying has given birth; of the conditions under, and the elements in, which their work is carried out, and the difficulties and dangers they have to encounter. Flying is essentially a profession for the younger generation. The strain is too great for men of more mature years. To withstand such strain requires all the vigour, the recklessness, the iron nerve of youth. It is a profession that offers an irresistible appeal to the healthy-minded, sport-loving youth of Great Britain, to whom adventure is the nectar of existence.

The writer's chief endeavour in the opening chapters has been to help the young man who wishes to adopt "Flying" as a profession. Part 2 of the book is composed of a collection of incidents taken from the diary of an air pilot on active service somewhere in the North of France. They are given in their original form. I also wish to thank the editors of the *Daily Mail, Daily Express, Daily Chronicle, Evening News,* and *Boys' Friend* for their courtesy in permitting me to use, in a few instances, material embodied in articles appearing in their journals.

E. C. M

London, 1917

Introduction

In writing of modern aviation it is to be regretted that the sport or science, call it what you will, was developed more in two years by the war than would have been possible in twenty-two years under normal conditions. Prior to 1914 we did not look upon aircraft and aviation with the degree of interest that their useful qualities warranted. Instead we were apt to regard them rather in the manner of a sporting spectacle, in much the same light as a football match, or a boxing entertainment, or as the *pièce de resistance* of the showmen; thus aircraft, the greatest and most potential discovery of all the ages, had to prove their worth in the maiming of humanity and the destruction of property.

Quietly and unobtrusively they were introduced into the plans of war; it must be admitted greatly despised and with a strong feeling of repugnance. Gradually—so gradually as almost to be unnoticed—they began to prove their worth.

From the very first days of the war it began to be realised that we must have aircraft. Our large navy was in desperate need of seaplanes to hunt the enemy warships from their lairs and his merchant- men from the seas. In the same way our tiny army required aeroplanes, but for a somewhat different reason: to be prepared against all enemy surprises, which in those desperate days of early 1914 would have been fatal.

As the war developed, the various belligerents began to settle down, to restore order from the chaos, and to prepare for a long and arduous campaign. Then the cry came for aircraft, more and more aircraft. In England the great engineering shops and factories were peremptorily ordered by the government to abandon their work and to construct aeroplanes as fast as they were able. Meanwhile the enemy, who had long been prepared, began to obtain an overwhelming mastery of the air—it will always remain a mystery why he did not use his aircraft to better effect at Mons and the Marne. After four and six months, fresh craft came out from England, and it was then the enemy, in his turn, was driven from the air. For some time we were allowed to retain that mastery, then the enemy came along with a rush with the new and powerful Albatross and Aviatik, and again we retired into the background for a time. Meanwhile, aeroplane factories were springing up all over the country, and the production of machines was going up by leaps and bounds; undeniable proof this of the value such craft were to the military commanders. Thus the mad race went on. Fast, graceful, single-seater scouts, slower and larger reconnaissance craft; huge, powerful-engined battle-planes made their appearance in quantities hitherto undreamt of, and were despatched in never-ending stream across the Channel, there to play their part in the war.

Dipping into the past, it may be said that by 1784 flight by balloon was well under way, and that year a woman—Madame Thible—made a trip in the presence of King Gustavus III of Sweden, that lasted three-quarters of an hour. She reached an altitude of 9000 feet. The following year the first cross-channel trip was made by Blanchard, with an American doctor named Jefferies for passenger, together with a supply of provisions and ballast. This weighed the balloon down to so great an extent that she almost sank into the sea a few moments after starting. Ballast was thrown overboard and she rose, only to sink down again. Hurriedly more ballast was dropped, but it had no effect, and was followed by everything on which the aeronauts could lay their hands, including provisions, books and a mass of correspondence. At last the French coast loomed into view, but the balloon was now sinking rapidly. The wings were thrown overboard, but that had no effect. The aeronauts commenced to strip themselves of their clothing. Then Jefferies proposed to jump over the side into the water, and was about to do so, when the balloon rose suddenly into the air, and they landed on the hills behind Calais.

Aircraft played a great part in the Franco-Prussian war, and during the siege of Paris alone as many as 66 balloons left the stricken city, carrying 60 pilots, 1 02 passengers, 409 carrier pigeons, 9 tons of letters and telegrams, and 6 dogs. Five of the dogs were sent back to Paris, but were lost and never heard of again, while 57 of the carrier pigeons carried 100,000 messages. Of the 66 balloons 58 got through, 5 fell into German hands, and 2 into the sea.

Among the more historical trips is that of Gaston Tissandier, who went over the German lines, and dropped 10,000 copies of a proclamation addressed to the soldiers, asking for peace, yet declaring that France would fight to the bitter end.

In South Africa an observation balloon was in use at Ladysmith for twenty-nine days, doing extremely useful work in spotting the Boer artillery. The pilot of an observation balloon reported the enemy's position on Spionkop to be impregnable, and, at Paardeberg, another disclosed the precise position of Cronje's force and directed our artillery fire thereon.

Of all the Great Powers, Italy is more responsible, perhaps, than any other for the evolution of aircraft. From the sixteenth century the most accomplished Italian scientists have given their attention to the solving of the riddle of the air. Such names as Leonardo da Vinci and Fausto Varanzio stand out prominently in the history of aviation; and today, (as at time of first publication), the Italian rigid airships are the best in the world. It was, however, mainly due to the efforts of two Frenchmen that prominence was first given to aircraft. Joseph and Stephen Montgolfier were the sons of a rich paper-maker of Annoney, and the story goes that, while rowing, Stephen's silk coat fell overboard into the water. When drying the coat it was noticed that the hot air tended to make it rise, and the upshot of the affair was the Montgolfier balloon. Since those days France has devoted herself almost entirely to the development of aeroplanes, which are second only to those of German manufacture. To the latter power honour, however unwilling, must be given as regards aircraft. On the outbreak of war her aeroplanes were the finest in the world, and her Zeppelins were beyond comparison. Great Britain possessed an advantageous lead in the matter of aeroplanes.

The development of aviation in this country was mainly due to the untiring efforts of the Royal Aero Club affiliated to the Féderation Aéronique International; and the splendid encouragement of the proprietors of the *Daily Mail*, who generously put aside an aggregate sum of £37,000 towards prize-money for aeronautical events. The Féderation Aéronique had already branches in America, Argentine, Austria, Belgium, Britain, Denmark, France, Germany, Holland, Hungary, Italy, Norway, Portugal, Russia, Spain, Sweden, Switzerland. In England the R.A.C. controlled every matter connected with aviation, such as the arranging and governing of competitions, the granting of pilots' certificates, and the ruling of the air. Up to August, 1914, they had already granted 926 certificates, of which 863 were aeroplane, 24 airship and 39 aeronaut (balloon).

The first of their competitions for the Britannia Challenge Trophy was carried off by Captain C. A. N. Longcroft, R.F.C., in 1913 with a flight from Montrose to Farnborough *via* Portsmouth, a distance in a direct line of 445 miles. It was the R.A.C. that arranged the *Daily Mail* competitions, several of which have yet to be carried out, including the £10,000 Cross-Atlantic (by aeroplane) . The *Daily Mail* International Cross-country flight for £1000 was won by Louis Blériot, July 25, 1909: it is needless to remark that this flight has now become an everyday occurrence. The £10,000 London to Manchester flight was awarded to Louis Paulhan (France). The second £10,000 circuit of Britain of 1010 miles was carried off by André Beaumont; and J. T. C. Brabazon was successful in the National *Daily Mail* £1000 for a flight of one mile in an All British machine.

The highest altitude that had been reached in Great Britain was 14,920 feet; the greatest distance flown 287 miles; and the longest duration 8 hours 23 minutes.

Whether we were prepared for the war is a matter for too extensive a discussion for this little book, but the fact remains that the number of firms engaged in the manufacturing of aeroplanes could be counted on both hands, and that we were without a useful and reliable engine of British construction.

CHAPTER 1

Joining the Service

The Air Service is young, very young; it is like an overgrown schoolboy, strong, healthy and full of life, but lacking just that sense of proportion that distinguishes the schoolboy from the man. It is wise, for it is endowed with the wisdom of initiative, courage and resource. Turned loose into an entirely novel and little understood element, it has had to create its own methods of procedure, its own ideals, its own traditions. Reference to the policies and the formulas of past generations are impossible, for there are none!

The main principles of aerial warfare are entirely new; in every combat, and in every raid, some precedent is established, some new form or theory of attack is set up. To the airman every day is alike. In times of peace he risks his neck as much as he does in time of war, save that engaged in the latter he has the additional unpleasantness of shell fire. He willingly gives all, but asks for nothing. He is the knight-errant of the twentieth century.

In days of the past, it was the cavalryman, wounded and galloping across country, with a hundred foemen hard at his heels, who first brought news of the enemy to the general in command. His was a pleasant occupation, that smacked largely of daring and romance. He stood an excellent chance of getting a bullet through his lungs, or of being clapped into an enemy prison. Today there comes flying across the heavens a resolute young hero, in a few feet of wood and fabric, throwing defiance to shot and shell alike, suspended thousands of feet up between heaven and earth, peering from that swaying aeroplane at the panorama of the earth beneath.

This is the age of science and invention. War on and over the earth, on and under the sea. For many years we have steadily been putting behind us the barbarities of our forbears, we have become more civilised, and, though more civilised, more barbarous. This is no paradox; science has made great and wonderful strides, but science has been more devilishly ingenious than any torture of Spanish Inquisition days.

The airmen who pilot their frail craft over hill and valley, sea and land, across cloud and through fog and mist, are the privateers of modern times; but for them there can be no capture, no quarter: only victory or a thousand feet drop to the cruel earth below. Through their young veins must flow the blood of a Drake, of a Philip Sidney, or Nelson. Theirs must be the courage of a conqueror, the heart of a lion, the nerve of a colossus.

No bounded ocean is their sea, but the infinity of space. The ship's compass is their best friend; for they manoeuvre their craft like a ship at sea. Wind and weather affect them as they would a mariner. For rock, shoal, sandbank and channel there are the high hills, the tall factory stack, the church steeple, and the deep valley. Landmarks there are, but always below, not on either side. Railways, roads, rivers, fields, woods and hills form the colour scheme of the surface of the earth, by which the air pilot steers a course.

This, the youngest and most important service, is essentially one *for* the young man and *of* the young man: a service the future of which is being steadily built up by the "muddied oafs and flannelled fools" of the playing-fields of the public schools of Great Britain.

Immediately after leaving school is the most perplexing period in a boy's life. Not only for the boy himself, but for his parents, for then has to be considered his future career. What is the boy capable of? What are his own personal wishes? What profession is he best adapted for physically? It is indeed a momentous question.

It is worse than useless for the boy fond of good, wholesome, out-of-door exercises and games to be put into an office or to study for the bar, or to mope his young life away pen-driving. And, on the other hand, it is a positive torture for the youth with distinct literary taste, or love of things scholastic, to take up a commission in one of the services, or to go in for farming or a similar profession.

Taking everything into consideration, at least eighty *per cent*, of boys may be grouped into the former class—that is to say, they wish to adopt a healthy, open-air profession; and for this type of youth nothing can be better, and nothing can offer greater inducements, than the profession of the airman. It is a calling that appeals irresistibly to a boy's heart.

The best possible training for the pilot of the air are outdoor sports and games. Football, which teaches the boy to keep his head in all emergencies, to keep his feelings always well under control, and to learn to obey implicitly the discipline of the referee's whistle will prove invaluable to him when learning to fly, when he will be subject to every kind and manner of unexpected and sudden mishap and accident.

Cricket will teach him patience, judgment—so invaluable when landing an aeroplane (which, incidentally, is by far the most difficult feat to accomplish in flying)—and a steady eye.

Swimming and running will develop those muscles of the back and thigh which are use extensively in the pilotage of the aeroplanes.

Again, the sensation of a horse jumping a hedge is exactly similar to that of an aeroplane just getting off from the ground. With skiing, on the other hand, there is the feeling—and, in fact, the action—of plunging desperately into what, at the first attempt, appears to be an interminable and awful space. This is exactly the feeling experienced by the novice in his first trip up aloft. There is a strong similarity to skiing at the moment that the nose of the machine is suddenly put down, and she commences to sink rapidly towards the earth.

The next matter to be taken into consideration is that of physical peculiarities. The would-be pilot must be neither too tall nor too short. This is essentially a matter to do with the steering of the aeroplane. If he is too tall, he will find himself very cramped in the confined space between the pilot-seat and the rudder-bar. If he is too short he will discover that his legs will not be long enough to reach that all-important adjunct.

Again with regard to weight, for preference he should be on the light side. There is not very much room in an aeroplane, and, for reasons with which we will deal, the machine is only capable of lifting up to a certain weight.

Take into consideration that an aeroplane is often required to take up two passengers, not to mention bombs, grenades, spare petrol and a machine-gun; every extra pound of weight is of the utmost importance.

His stomach must be strong, for with a weak stomach he will be liable to air-sickness.

Further, he must be possessed of good health. He must not suffer from heart trouble. It has been proved by several very eminent doctors that the rise and the descent through the various altitudes of the atmosphere affect the heart greatly.

Again, he must have good eyesight. This is imperative, for the best part of his work will take place at an altitude of ten thousand feet above the earth. The best age for an air pilot is between nineteen and twenty-four.

The life of a pilot—that is to say, his flying life—varies from three to five years; I may say eighteen months under war conditions. Never more. The great strain on the nerves, although not felt at the time, begins to make itself apparent after two years of flying; then the pilot discovers that he is no longer so keen on going up as he was, that he gets "cold feet" more frequently than he was wont to do in the early days, that he has no longer the nerve to do the little tricks, upon the performance of which he formerly prided himself.

A good air-pilot must be born so, he cannot be made. After years of experience a man may become expert in trick flying, landing, getting off, etc.; but, however long and however diligently he may strive, he can never become the equal of the natural pilot.

Before applying for a commission in either service the aspirant to flying honours must first decide which of the two branches he wishes to take up. The two branches, by the way, are pilotage and observation. The difference between the two I will here briefly endeavour to explain.

The pilot is concerned with the flying of the machine, the care of the engine, spare parts, etc., and is responsible for the general condition of the craft; also to see that it is properly tested before each flight.

On the other hand, the observer has a great many subjects to learn. He must be at one and the same time wireless expert, gunner, rifle-shot, artist, photographer and map-maker. He must know something about heavy artillery.

The observer in the Royal Flying Corps is given equal rank to the pilot, but can only wear a half-wing on his tunic where the pilot has full wings.

In the Royal Naval Air Service observers are permitted to wear the bird on their sleeve immediately on joining. However, they are of different rank from the pilot, being either lieutenants or sub-lieutenants, Royal Naval Volunteer Reserve.

Chapter 2

The Airman's First Days

The appointment to a commission in one of the flying services can be either temporary or permanent. The former holds good until the end of the war, the latter for as long as the would-be airman wishes to retain it. For a period of from four to six months he must undergo a probationary course; if after that time he has served satisfactorily he will be confirmed in his rank.

Upon first joining up he will receive a uniform allowance of £20, and at the confirmation a further £20. These amounts should easily cover his requirements and enable him to buy a complete flying outfit. During the probationary period he will receive 14s. a day in pay; when he is confirmed in rank, 18s. a day in the Royal Naval Air Service, and 20s. per day in the Royal Flying Corps.

Service etiquette plays a prominent part in the matter of uniform. In the military wing he will be expected to wear the button-over tunic and forage cap of the Flying Corps, with breeches and long brown field-boots.

In the R.N.A.S. the matter of dress is a more difficult and more delicate one In the first place, with regard to the cap, there are four entirely separate badges in the Naval Service: they are (1) the big silver anchor and the gold crown of the regular Navy; (2) the smaller replica of the Royal Naval Reserve; and of the Royal Naval Volunteer Reserve, to which latter branch the aeroplane observer always belongs; and lastly the silver bird of the R.N.A.S., worn only by pilots.

In hosiery the naval flying man must confine his taste to plain white shirts with collars to match; black ties, and socks of the plain black variety. His shoes must be unadorned of toecap, and it is a cardinal sin to leave the buttons of his jacket undone, if he reveal as much as a button of the waistcoat beneath.

There is an amusing story told concerning a famous English airman who has since resigned from the R.N.A.S. On the occasion of his appointment to the service he had to visit a certain big man at the Admiralty, and arrived there in the brass hat of a full-blown naval commander, with a black and white striped tie, in which there coyly reposed a large diamond pin.

When the interview was over the big man called him back.

"You've forgotten something."

"What is it, sir?" the airman inquired.

"Your pink shirt and your purple socks," was the reply.

Another new hand—an Australian—presented himself to the astonished and apoplectic commanding officer of his first station wearing a blue monkey-jacket, white flannel trousers, green socks, and brown shoes.

Luckily he was a good-tempered youth, or he would never have been able to live down the subsequent ragging he got from all the other members of his mess.

Flying-clothes must be the warmest procurable: a black or brown leather coat lined with lamb's wool, with trousers to match. Good flying-coats cost from three to five guineas, and the trousers range from a guinea to thirty shillings in price.

A khaki balaclava helmet, a wool-lined aviation cap fitting closely round the skull, and costing approximately half-a-guinea. A pair of triplex glass goggles, price 12s. 6d.—cheaper ones of ordinary glass can be obtained as cheap as 3s. 6d—but it is always advisable to get triplex, as in the event of a smash-up ordinary glass would splinter, fly into the eyes and possibly blind one for life.

A good pair of leather gauntlets, large enough in size to permit the wearing of a warmer pair of woollen gloves beneath, and a grey sweater to wear underneath the leather coat are all that are required, bringing the total cost to about £6.

As in other professions and walks in life, a certain slang has sprung into being in flying circles, and this the new hand will discover will take him a considerable time to pick up—at least, with any degree of satisfaction or success.

First he will discover that a "quirk" or a "hun" is no less a person than a youngster who aspires to flying honours, and who has not yet taken his ticket. Even the aeroplanes themselves have nicknames, as the "Bristol Bullet," so called because of its peculiar shape.

Airships and balloons are always referred to—and somewhat contemptuously, it must be admitted—by aeroplane pilots as "gasbags." The small, silver-coloured airships that are to be seen occasionally floating over a certain western suburb of London are known in the Service as "Babies," on account of their diminutive size; on the other hand as "Blimps," and again as "S.S.'s"—submarine seekers—that being their principal duty when on active service.

Various parts of the machine have their own particular nickname, as the "fuselage," or body which contains the engine, pilot and observer's seats, and the petrol tanks. That wonderful control lever which is placed immediately before the pilot's seat in the fuselage, and which manoeuvres the machine both upwards and downwards, and to the left and to the right, or, in the terms used by the R.N.A.S., to port and to starboard, is known as the "joy-stick." No self-respecting pilot will ever refer to a trip in the air as such, but rather as a "joy-ride." A bomb-dropping expedition or a raid he speaks of as a "stunt."

To "nose-dive" is for the front portion of the machine to plunge suddenly downwards at an angle of approximately ninety degrees with the earth. To "pancake," the aeroplane must fall flat to the earth. It is possible sometimes to recover from a "nose-dive," but never from a "pancake." Sometimes in banking—turning in mid-air—a pilot will overdo the angle at which he turns; the result is that the machine commences to rotate, and whirls round like a humming-top; this, again, invariably develops into a "nose dive," and is known as a "spin."

The majority of pilots, when first starting off, run their machines some distance across the aerodrome, then rise gradually at an angle of about fifteen degrees with the earth; others, on the other hand, prefer to run their machine a considerably greater distance across the ground, and, thus attaining a much greater speed, to rise almost vertically for about two hundred feet, then to flatten out and bring the machine level: this trick is know as "zumming."

To "switchback" is to fly up and down, up and down, as the name implies.

Immediately after leaving the ground the plane invariably commences to plunge and to dive like a ship in a stormy sea—this is when it enters a patch of rarefied air known as a "bump;" this latter often causes the machine to drop suddenly, and drops of as much as two hundred feet at a time have been recorded.

No airman is capable of talking through his hat—at least, not literally, for he does not possess such a thing, that article of his attire always being referred to as a "gadget."

To have "cold feet" in the air is to have a bad attack of nerves or funk. One day at Hendon, before the war, a well-meaning but somewhat dense journalist attached to a big London daily was told Hamel was suffering from "cold feet."

Imagining that "cold feet" meant some ailment of the feet, like chilblains, and solicitous for his welfare, this enterprising individual approached the famous airman immediately after his descent from a trip up above.

"Excuse me asking, but is it true that you suffer from cold feet, Mr. Hamel?" he asked.

Hamel's reply is not recorded.

CHAPTER 3

The Initial Flight

Once in the Service, the R.N.A.S man may be selected for one of three branches of flying, namely, seaplane, aeroplane—which, incidentally, is far preferable to any other branch, and holds forth more opportunities of active service and kite balloon, probably the safest and most comfortable job of the war, but dull—deadly dull.

For the sake of those of my readers who do not know of the captive kite-balloon, I will here briefly explain. It is a queer sausage-shaped craft, that is tethered to a steam-winch on the ground somewhere beneath it by means of a stout steel cable. Usually situated some five or six miles behind the firing-line, the basket of the balloon will only hold two observers at one time. It is connected to the big guns by telephone, and is useful for the direction of artillery fire, which it does by telling the men at the guns whether their shells are falling over, under, or to the left or right of the target that they are aiming at.

The first day in the life of the "new hand" at the Service school is not always the pleasantest of memories. He discovers that, from a man of parts, he has suddenly been converted into a very junior sub, and is at the beck and call of every member of the mess, with as much or more gold braid on the sleeve of their uniform.

For the first few days he is allowed to wander round at his own sweet will, in order to get the hang of things. To him the matter of greatest importance are the machines, for very often he has never even seen an aeroplane at close quarters, and should he be foolish enough to ask absurd questions, he will always find someone ready with a fitting answer.

He will be told wondrous stories of the time the machines will remain in the air, the breakneck speed at which they will travel, and of the enormous height to which they will climb.

The next most important thing to the actual flying is a thorough knowledge of wireless telegraphy, for without a wireless instrument on board an aeroplane is little better than useless to the army in the field; and, having got the wireless set on board, the pilot or the observer—whosesoever duty it is—must be able to send messages, clearly and distinctly, on the Morse key.

A good tip to the youngster thinking of taking up flying for a profession is to buy a copy of the for the sum of 5s. 6d.), and to teach himself to read by sound.

In service circles the dot and the dash of the Morse code are known as "iddy" and "umpty," respectively. It is a simple matter to learn to send and to receive wireless signals; but to know how to erect and dismantle a wireless set, and to have a sound knowledge of the theory and the working of the thing, and to be able to take to pieces or to repair at a moment's notice, any portion of the instrument that may get out of order, is a more difficult matter.

That requires several months to acquire, but the "Quirk" will be given a useful, though somewhat "short," course under an expert wireless operator before he is expected to know these things.

At last the great day arrives when he goes for his first trip up aloft. After donning a leather coat, and trousers to match, a skull cap and goggles, he is ready for the fray, and sits himself gingerly beside what at the first seems to him to be a particularly violent and a particularly ill-disposed individual with a simple wonderful flow of language, an instructor in a "box-kite." Then the engine is set going.

The instructor bawls some remark into his ear, which, for the life of him, he cannot catch. A long and rapid journey across the bumpy ground, a weird sensation of rising into space and he is up in the air at last. Then the machine gets into the "bumps;" she dips, and drops, and sways, first to one side and then to the other, until the poor unfortunate individual begins to wonder if he will ever get safely to the ground again.

There is a pandemonium of noise. The wind rushes by his face at an alarming rate. He feels himself perspiring all over, and particularly in the palms of his hands. He grips the nearest available object, as a drowning man would clutch at a straw. With every fresh plunge and dip he increases that grip.

The instructor shouts at him at the top of his voice, but he hears nothing; only the racing engine and the whistle of the wind. And then for the first time he ventures to look over the side. Could that curiously-scattered collection of pigmy buildings, long, ribbon-like roads, and distant, narrow, gleaming line of railway line be the earth?

He decides that it is, and is at last beginning to feel comfortable, when the machine begins to heel over violently; it is the worst shock that he has yet had. He grips with both hands as tight as he is able, shuts his eyes, and waits for the worst. By the time his eyes are open again the machine—by what seems to him to have been a miracle— has righted itself and is flying smoothly through the air. Never before has the world appeared so beautiful nor so diminutive in size.

For another five minutes or so the instructor flies to and fro above the aerodrome, then down goes the machine, much to the astonishment and alarm of the bewildered "quirk," who suddenly finds the earth rushing up to meet him. How he fears that moment when a landing must be made, and how relieved he feels when he realises there is nothing in it in the least degree terrifying.

Very gently the aeroplane skims on to the landing-ground, like a seagull lighting in the crest of a wave, and all is over; he is safe back again on Mother Earth. Silent and subdued, he clambers out of the aeroplane. How did he enjoy it? "Very much indeed," he answers in a husky whisper, and the instructor turns his head away and smiles. He has taken "quirks" up before.

CHAPTER 4

The Perils of the Air

For the first few trips up aloft the beginner is always accompanied by an instructor. First he is taken up as a passenger, and his only duty is to sit in the observer's seat and do nothing. Then gradually he is allowed to fly the machine himself. This he does in a double-control—that is to say, an aeroplane with two sets of controls, one of which the instructor makes use of and the other is in his own hands.

He is taught that every movement of the control must be slow and gentle, otherwise the machine is sure to lose its stability—balance—and go crashing to the ground below; that an inch too much with the rudder-bar will invariably mean a "spin," or a too jerky movement on the control- bar a "pancake "or a "nose-dive."

Getting off from the ground is a comparatively simple matter; but the moment of first entering the air is the most dangerous and trying of all. Should the engine fail, the chances are a hundred to one that the machine will crash into a hedge, or a tree, or land in a valley. The "bumps" are most frequent over houses and buildings, and particularly so on a dull morning, when the sun is breaking through the clouds, which send the craft plunging and tossing in all directions. This is the test that will show if a man is a good pilot or no.

Once clear of the "bumps," the first thing to be done is to get "height." With a ship at sea the safest sailing is in mid-ocean, far from the land. In a similar manner, the greater the altitude the safer is the flying.

When near the ground, the air pilot has very little choice in landing-places and very little time to prepare for a landing. The higher up he is, the greater range of country he has to choose from, and the more time he has to regain control of his machine.

At a rough estimate, one may say that at a height of 500 feet he has only an area of a square half-mile to land in; at 1000 a mile; 2000 two miles; 5000 five miles; 10,000 ten miles, and so forth. Some few months ago a pilot at Brooklands flew up to a height of about 15,000 feet, shut his engine dead off, and glided down into Hendon aerodrome a distance of just over twenty miles.

Having got clear of the "bumps," the next danger is the clouds, which have a very strange effect on the stability of the craft. They should always be avoided when possible. Fog is a very terrible element to encounter in mid-air, and the sensation of being fog-bound is the worst that the human brain can conceive. Nothing in sight, with the blinding fog on either side, and not knowing any moment that he will not be colliding with some high points of the earth, the air-pilot positively dreads the fog.

The writer remembers well the case of an airman fog-bound last winter at an aerodrome near London. For two hours he was flying up and down, up and down, over the aerodrome, without being able to find it. The spectators on the ground could hear the hum of his engine distinctly, but could not see him, and neither could he see them. Eventually, with the aid of landing-flares and Verey's lights, he was able to land; but for weeks afterwards was a nervous wreck, and could not fly again for nearly a month.

After several trips with the instructor, and having satisfied that individual that he has gained sufficient knowledge of flying, the "quirk "is allowed to take up a machine by himself.

At first he flies it up and down, over the aerodrome, then gradually gets on to left and right hand turns, and then to landing the machine.

Now, landing is the most difficult feat of all in flying; it requires both good judgment and good nerves. Before landing the pilot must discover the direction of the prevailing wind. This he can do by watching the smoke of a high chimney, or of the locomotive of a railway train. Having discovered the direction of the wind, he must land dead against it, otherwise the machine will be caught in a sudden gust and toppled over.

For a day or two he will be kept on "landing" practice, and then he will be allowed to try for the Royal Aero Club aeroplane certificate. The tests and conditions for this are as follows: The candidate must be over eighteen years of age, and of British nationality; he must accomplish the three following tests, each being a separate flight

A and B.—Two distance flights, consisting of at least five kilometres (three miles, 185 yards) each in a closed circuit.

C.—One altitude flight, during which a height of at least 100 metres (328 feet) above the point of departure must be attained, the descent to be made from that height with the motor cut off. The landing must be made in view of the observers, without re-starting the motor.

The candidate must be alone in the aircraft during the three tests.

Starting from and alighting on the water is only permitted in one of the tests, A and B. The course on which the aviator accomplishes tests A and B must be marked out by two posts or buoys, situated not more than 500 metres (547 yards) apart. The turns round the posts or buoys must be made alternately to the right and to the left, so that the flight will consist of an uninterrupted series of figures of 8.

The distance flown shall be reckoned as if in a straight line between the two posts or buoys. The alighting after the two distance flights in tests A and B shall be made (*a*) by stopping the motor at or before the moment of touching the ground or water; (*b*) by bringing the aircraft to rest not more than 50 metres (164 feet) from a point indicated previously by the candidate.

The decision of the committee of the Royal Aero Club in all matters connected with the test is final, and without appeal.

The certificate itself, which is a handsome, leather-bound affair, in the shape of a pocket-book, can be obtained by sending along the certificate of the flights accomplished, together with £1 1*s*., a photograph of the applicant, particulars as to birth, etc., to the Secretary, Royal Aero Club, 166 Piccadilly, London, W.

His "ticket" having been obtained, the "quirk"—who, incidentally, is now a "quirk" no longer—is given a little more practice in flying slow machines, in order to gain confidence, and is then sent on to his first war station to learn to fly the faster battle-planes and war machines, and at the same time is confirmed in his rank.

Even now his flying education is by no means finished. After learning to fly the faster machines, he will be put through a course of bomb-dropping. After that a spell of cross-country work will occupy his time; learning to fly from above by the position of landmarks, roads, rivers, railways, etc.

After this he learns to steer a course by compass, gets practice in machine-gun firing and dissembling while in mid-air, and then he is ready at last for the great adventure across the water. One fine morning he will set out on a brand-new war-machine for somewhere in the north of France.

CHAPTER 5

The Spirit of the Air

The great war has brought in its trail horrors innumerable, but, as if in compensation, has brought to light all that is best in our men.

The heroism and courage of the airmen were without precedent, but none the less admirable. Those stripling pilots of the air that flew undaunted over shell-fire in all weathers and at all times have opened up a chapter in our history that nothing can rival.

Who can define the psychology of these young men who can meet death as an old acquaintance and pass him, mocking, by—who laugh at fear, and make a jest of danger? Is it that they are without nerve entirely, or is it rather a pose, a lovable bravado that hides their true feelings? Is it that they are rather less devoid of fear than their brothers in the trenches? Hardly. We have known them, you and I, reader, in the last few years, but under a different guise—as happy, laughing schoolboys, as young men plunging into life, the "flannelled fools and muddied oafs" of Britain, and suddenly they have become men, ready and eager to share a man's burdens and responsibilities, yet no whit altered; but deadly in earnest when there is work to be done on the other side.

Undoubtedly the air does affect a man to a degree, and endows him with that strange malady, flying temperament, that makes him reckless, and, to a certain extent, headstrong; occasionally to get out of hand, and to find rules and discipline chafing and irksome. But then the air has a call of its own that few can resist; that runs through a man's veins like flame, and whispers courage and defiance into his ear, that invites his sympathy, his love, his esteem. But the air is a fickle mistress, and woe betide he who dares to slight her or make free at her expense; he must pay the penalty, and that penalty is—death.

Every known sensation is experienced in flying: joy—the joy of youth astride the dull old world, accomplishing what previous generations dared not to attempt; excitement, to feel the cool air brushing one's cheek, and whistling past one's ears; fear, danger, hope and despair; all are crowded into this one brief hour of life.

Day after day, in all kinds of weather, the airman must go up, for the battle seldom slackens and never pauses on the earth beneath. One day reconnoitring—that is, making a long flight over the enemy's country under a continual bombardment from the Hun anti-aircraft guns, noting any fresh movements of enemy troops, gun emplacements, headquarters, supply depots, ammunition columns, or any unusual activity on his roads or railways. Another day taking part in a bombing raid on some distant military centre, or perhaps out fighting enemy aircraft; but always taking his life in his hands, and never knowing each morning as he sets out whether he will return again.

It is the proud and honest boast of the British Air Services that they never advertise; and what we lack in that respect, our enemy make up for. We have our Immelmanns and our Boelkes, but their identities are hidden under the simple pseudonyms of Lieutenant X—— and Lieutenant Y——. They perform their daring feats, not for their own vainglory, not for the sake of decorations, but from keen sense of duty, love of their work, and for the further honour of the famous corps of which they are units. It is this policy of eternal silence that has so completely shattered the moral of the German airmen in Flanders, and driven them almost entirely from the air.

In many ways the air is own cousin to the sea, for there is a chivalry of the sea which has been a tradition for tens of centuries; a freemasonry of good feeling and sportsmanship among those who have their business in great waters.

The chivalry of the air is none, the less real because it has no traditions to fall back upon. Nature herself has made the man of the sea and the man of the air sportsmen alike; has given them an instinct for "doing the right thing."

The Air Service has, in addition, a quality exclusively its own; I mean its youth. It is just like a healthy schoolboy, intensely alive, active, happy-go-lucky, yet ingenious enough where matters of technique are concerned, and always eager to be out for adventure.

But it is just these tremendous dangers which are the breath of life to this splendid schoolboy (even in age he is often little more). There is a sporting touch in this ceaseless duel with fate, in this juggling with life and death. That touch is transmitted to the less figurative duels when there is a tussle in mid-air with a flying Hun, when it is his life or yours.

On second thought I withdraw that word Hun in relation to the German airman; I continue to apply it with all the vehemence I can muster to the crews of a baby-killer Zeppelin, but one's adversary in Albatross or Halberstadt is an adversary worthy of the name. Here, almost alone in all phases of modern warfare, remains the personal touch. Up there in the awful solitude of space two human beings pit their brains and courage one against the other, with death each moment before the eyes of both. It is a strange turn of things that the latest development of modern science has brought about a revival of mediaeval chivalry, the single combat.

I have mentioned the freemasonry of the air. Any airman who has seen any fighting could give you countless instances of it. Your German air man treats you as an honourable foe, and you treat him as one. That constantly recurring phrase, "An aeroplane was forced to descend and its two occupants taken prisoners," means that those prisoners, whether Germans or English, were treated honourably, even ceremoniously. A wounded aviator landing in the enemy's lines is lifted from his seat with every care, and is almost invariably saluted. I have known on five separate occasions airmen fly over the enemy simply to drop the personal belongings and effects of the men whom, in a terrific mid-air struggle, they have succeeded in sending crashing to earth and death. German airmen have done the same, and seen to it that his comrades should receive the cigarette case or bundle of personal papers of a fallen foe.

One of the most dramatic incidents of this drab war was the dropping of a wreath from an English aeroplane in honour of the dead hero of the German Air Service, Immelmann.

An airman likes an opponent worthy of his mettle; he likes even chances and the prospect of a good fight. I shall always remember the disgust at a certain war aerodrome recently. The approach of a Zepp had been reported, and all was excitement. Aeroplanes were dragged from their hangars, and off they went at lightning speed. Soon the return. Disgust was on every one's face. "We thought there was going to be some real fun," was the general grumble. "Zepp? Not a bit of it; only a sausage balloon."

Danger the airman shares with the soldier in the trenches. Many a tale could be told of the awful deaths, of roasting when the machine catches fire, of hours of agony with a shattered leg or arm when, at all costs, the machine must be piloted to safety and a life (that of the observer) saved. But such things are the lot of most men who fight. It is the cheery sportsmanship, the good fellowship, the national instinct to fight and behave like a gentle- man, that have become characteristics of airmen of all nations, which I have tried to emphasise.

Such is "playing the game" in the Air Service. Often it is a cheery life, but it is always a trying one.

Chapter 6

Seaplanes

The seaplane, as its name implies, is used solely for flying over tracts of water. It is identical in shape with the aeroplane, but with minor variations. It is considerably heavier than the aeroplane in weight, and is more of the formation of the boat, though following the same "streamline" principles as the aeroplane.

The engine-power varies from 70 to 150 horse-power, but the machine is much slower in transit and in climbing even than several of the lesser horse-power land machines. The fuselage, or body, is like a flat-bottomed boat, in the bows of which are the engine and the propeller. Immediately in the rear of the engine are the pilot's and observer's seats, side by side, and not, as in the aeroplane, the one behind the other. Again, in place of the wheels of the landing chassis of the aeroplane are two boat-shaped floats; these are hollow in formation, very heavy, and extremely fragile. When landing the seaplane on a rough sea, the part of the machine most liable to break up is the float.

With regard to the actual flying of the craft, where a mere touch of the control is capable of manoeuvring the aeroplane up from the ground, it requires the grip of a Sandow's developer to lift a heavy seaplane off the surface of the sea. Similarly, while manoeuvring in the air, the movements must always be of the gentlest nature, considerable muscular force is required to bank (turn) and climb the seaplane.

Landing is the most difficult and delicate manoeuvre in flying; it is a tricky performance to land an aeroplane, but it is doubly so to land a seaplane. Should the surface of the sea be the least bit choppy or rough, there is a grave risk of the floats breaking open, and the machine turning turtle, or diving down through the sea and precipitating the pilot to a watery grave.

The work of the seaplane may be placed in two categories: first, work from the shore, when a landing-station, bordering on the sea, is used as a base; and, secondly, flying at sea, when the craft is taken out on board a parent vessel, and flights are commenced from the middle of the ocean. With regard to the former, the work is for the most part of a defensive nature, as that of driving off invading enemy craft, and patrolling the coasts for enemy submarines. The work at sea is principally scouting for fleets, for a seaplane observer, at an altitude of 5000 feet, has a range of view ten times greater than the lookout man of any battleship or cruiser.

In this latter case, flights are usually terminated and commenced from the sea surface, alongside the parent ship; and when the craft are no longer in use they are lifted on board by means of a large crane and stowed away on a specially constructed deck.

From the point of view of interest, aeroplane work is preferable to that of the seaplane. Nothing more boring and dreary can be imagined than a long flight over an interminable stretch of blue water; the aeroplane pilot does, at least, have an ever-changing contour of hills and valleys, rivers and woods, towns and villages beneath him, whereas the seaplane man's view is confined to sea, sky and horizon, with perhaps an occasional passing ship.

One seaplane pilot of my acquaintance, in order to relieve the monotony, always took his dog, a staid and wise-looking Scotch terrier, with him. That dog can lay claim to holding the record among dogs of the world, for he has now flown considerably over 2000 miles. His method of aviation is peculiarly his own, for, once the machine has started and got under way, he curls himself up in the body of the fuselage and goes into a sound sleep, from which he does not wake until the engine stops again.

Seaplane flying in these days is beset with dangers of many kinds.

As an example, I will attempt to portray the average day's work of a seaplane pilot on active service, somewhere in the North Sea.

A scene of unusual activity is revealed by the breaking dawn, lat. "X," long. "Y." The sea is calm, the rising sun giving it that peculiar greyish-green tint, over which the early morning mist hangs like a pall. Through the mist can be seen the hazy, blurred outlines of the Fleet: squat, lumpy monitors, slim and graceful cruisers, sharp-nosed destroyers, submarines that hang, as it were, on to the surface of the water. Great towering battleships, dignified and stately, look down upon the smaller fry with apparent disdain. Far in the rear there is what at first appears to be an ordinary smug-funnelled tramp steamer; but a glimpse of the huge crane and queer, elongated shapes along her decks reveals the seaplane carrier.

Four o'clock in the morning. Though it is summer, the weather is cold and raw, the chilly breeze bites knife-like through one's clothes, fingers are all thumbs—rather a disillusion of the joys of flying. The engine stops, and coughs and splutters as if in protest at this extraordinary behaviour. Compass, maps, instruments are missing; the petrol tanks are unfilled, or the oil has been forgotten.

At last, creaking and groaning, the crane is lowered, and fixed to the craft. A few hoarse commands, and she is swung off the deck and dropped gently on to the sea, and off she goes, bound on a reconnaissance trip or target-registering. First taxiing far across the open sea, clear of the fleet. What a delightful sensation this is, skimming the water like a seagull, dipping and bowing gracefully; but it is quite another story when the sea is rough, and the swell threatens every moment to break up the floats and submerge the craft. At last up into the air, 200, 300, 500, 1000 feet, circling round the now, seemingly, stationary Fleet; how still and quiet they appear down below there!

The seaplane is usually a much slower craft to climb than the aeroplane, and some time elapses before a decent altitude is reached. The observer busies himself plotting out the course, testing the wireless gear, and preparing his report.

Scouting is the object of the flight, and scouting implies, for the most part, keeping a weather eye open for suspicious craft, enemy battleships, cruisers, destroyers and enemy submarines, the latter more easily distinguishable from a height, when the bed of the sea in the more shallow portions can be read like an open book, sandbanks standing out most prominently from the surrounding azure blue.

Target-registering, on the other hand, consists of following, or rather attempting to follow, a damnably perverse raft, on which a large target is lashed, at which the heavy guns of the Fleet are firing from a distance of from fifteen to twenty miles, and the observer wirelessing back the results of their attempts, also entering the same in his report.

To the uninitiated this report would at first sight appear slightly less understandable than a Chinese love letter or a Greek play. It is divided into columns; first there is the time of the entry, next the height at which the machine was flying, the approximate position, and, last, the nature of the observation. For example: 11.55 a. m. 6000 feet. Lat. 90, long. 70. 6. Large two-funnelled steamers, apparently merchantmen, observed proceeding in a south-westerly direction.

If the matter is of an urgent nature it is sent back to the fleet immediately by wireless, surely the most valuable asset to aviation that exists, and without which aerial scouting and reconnaissance work would be almost useless. The apparatus is light and extremely compact, consisting of one or two Morse keys and an aerial. The range of action—that is to say, the distance that a message can be either sent or received—is not very great, but such as it is, is invaluable. In a word, wireless in the navy is as near perfection as it is possible for a new science to be.

The observer makes a sudden movement with his hand in a south-westerly direction. Far down on the distant horizon is the long black sleuth-like form of an enemy destroyer. The wireless is soon busy ticking the gladsome news to the Fleet, now far in the rear. More and yet more black shapes appear, and then our own destroyers come up, dashing through the sea, at well over thirty knots an hour, leaving a line of churning white foam in their track. The enemy catch sight of them and then turn north at full pelt, our own in hot pursuit, until for fear of floating mines—it is a favourite trick of the Hun sportsman, when pursued to drop mines behind him, in the hope that they will strike the enemy ships—our own destroyers come back crestfallen and downhearted.

En passant it may be said that a seaplane battle is very similar to a fight between two aeroplanes, though usually more slowly fought out, and hence longer in duration. Such feats as "looping" or sudden nose-dives are generally impossible.

The morning's work is now completed, the recall signal is received via the wireless, and the great bird turns for home, not, however, without sighting several merchantmen and something which appears to be the periscope of a German submarine, but which, however, proves on closer inspection to be floating wreckage.

The British Fleet comes nearer into view, first the different shapes and sizes of the varying craft, then the funnels, then the masts, the rigging and the crew aboard. Throttling down his engine, the pilot sinks gradually lower and lower, and lands on the smooth surface of the water—strange to say a more difficult and more tricky feat than to come down on the solid earth for reasons too numerous to mention in this short chapter.

Another long "taxi" across the water to the side of the seaplane carrier, the creaking crane comes sliding out again, is fixed to the craft, which is hauled aboard, and stowed away until further required.

CHAPTER 7

A Zeppelin Chase

"X or Y airships participated in the attack on Great Britain last night; Z raiders were brought down." Hard official words these, that, read in the cold black and white of print, fail entirely to bring to the reader's mind a true sense of the danger and the nerve-racking conditions under which this novel form of warfare is fought out.

Let us imagine, if we can, the difficulties the aeroplane pilot has to face. It is dark—pitch dark—sky and earth are alike indistinguishable. Flying at the best of times contains a more than comfortable element of danger, and in the darkness this danger is accentuated. The darkness deprives the air pilot of all sense of direction and of locality, greatly hampers him in the manoeuvring of his craft, and renders unpleasantly possible a collision with another aeroplane on similar errand bent.

Starting out, there are a hundred and one small details to be attended to, as the testing of the engine, the trying of elevators and ailerons, and the examination of the petrol and oil tanks, in order to ascertain if there is a sufficiency of both to last a two or three hour trip. All this to be performed in the dark, with the engine screeching loud, so that a man may not hear a word, and the attendant mechanics indistinguishable in the gloom.

Fortunately for the pilot, a small dry-cell electric lighting set is installed in the body of every machine, and by this means the pilot is able to distinguish his instruments—a most necessary adjunct to safe flying—the altimeter, which records the height, "revmeter" which indicates the speed of the engine and the compass, more necessary than any other instrument for night flying.

Getting off from the ground is by no means a pleasant sensation. There are hangars, high roofs, and chimney-stacks waiting to be collided with, patches of thin and rarefied air, which will bump the machine down as much as thirty feet at a time; the ever present danger of engine failure, necessitating a descent to the darkened earth beneath, always so full of death-traps for the airman and his craft.

Clear of the earth, at about 1000 feet, there are, here and there, faint patches of light of dark grey and the subdued reddish glow of the distant metropolis; the locomotive of a passing passenger train, bright as a searchlight for a brief moment, then passing away into the outer darkness. Higher and yet higher; and the sensation! The mind of a Jules Verne or of an H. G. Wells could not imagine a feeling more eerie, more strange than this. Noise and darkness, the incessant deafening purr of the engine, the pitch blackness on all sides, relieved by the one tiny light inside the fuselage, as welcome and cheery to the airman as a distant lighthouse to a sailor in a storm.

Then the searchlights begin to blaze, creeping up across the sky in ribbons of shining brightness. One plays for a moment on the machine, the pilot is almost blinded before it passes on its strange search across the heavens. But a stringent search reveals—nothing! For an encounter with the raiding airship is not at all probable at an altitude of below 6000 ft., and from that height up to 15,000 ft; the only likely encounter is with the observation car of a Zepp. This car is usually suspended hundreds of feet beneath the mother-craft by means of a stout aluminium cable or cables, is about 7ft. by 5ft., composed entirely of aluminium, and contains sufficient space for one observer, who is in telephonic communication with the commander.

At last the pilot of the aeroplane has an instinctive feeling that a Zeppelin is somewhere near him. He cannot hear because of the noise of his own engine, and he cannot see because of the intensity of the darkness all around him.

The combat between the aeroplane and the Zeppelin might be compared to that between the British destroyers and the German Dreadnoughts in the Jutland battle. Dashing in with great rapidity and skill, the tiny one-gunned aeroplane fires its broadside, then makes off as fast as possible to get out of range of the comparatively heavy-armed airship. From thence onwards it develops into a fight for the upper position, for once above the Zeppelin the aeroplane pilot can use his bombs, which are considerably more effective than a machine-gun, and the broad back of the gasbag offers a target which can hardly be missed.

In manoeuvring, the aeroplane has the great advantage of being remarkably quick, both in turning, climbing, and coming down, whereas the Zeppelin again is a slow and clumsy beast at the best of times. The Zeppelin is very susceptible to flame and explosion of any kind; the gas in the envelope, a mixture of hydrogen and air, forms an extremely explosive mixture. The aeroplane, owing to the fabric of which it is composed, and the petrol needed for propulsion, is to a certain degree inflammable, but not nearly to the same extent as the airship. On the other hand, the airship possesses a distinct advantage in that it is able to shut off its engines, and to hover, which it is impossible for an aeroplane to do.

Again, in the matter of speed in a forward direction, and, for that matter, backwards also—for the Zeppelin engines are reversible—the aeroplane holds the palm with an average speed of sixty miles per hour, while that of the airship is only fifty.

The combat finished, the aeroplane pilot has yet to make a landing, surely the most dangerous and tricky manoeuvre of the whole flight. The difficulties and dangers thus encountered are too obvious to need explanation further than to say that the landing has to be affected in the dark, with only a blinding, dazzling, electric ground-light for guidance.

CHAPTER 8

The Complete Airman

The British Air Service is now a great army, 80 *per cent*, of whom, before the war, had never even seen an aeroplane, much less been up in one,—bank clerks, young merchants, undergrads, doctors, lawyers, journalists, all endowed with two sterling qualities required by the pilot of the air, courage and level-headedness. And how has this great miracle been accomplished? August 1914 found us lamentably short of both *personnel* and material, but what little there was of the very best. The already experienced pilots set to work with a will upon the more than generous quantity of raw material that came to hand. Within a few months their influence made itself felt. They taught the "quirk"—the airman's pet name for the novice—in their own simple and undemonstrative manner, that the air is to be respected, but never feared, the aeroplane treated as a being of life and animation, with quaint humours peculiarly its own, and not as a lifeless mass of metal and woodwork. Within six months the number of fully trained British pilots had trebled itself; within one the number had grown beyond all proportion, and still it goes on.

The usual method of training a new hand is to get him used to the air, which, though apparently harmless and void, is as tricky and treacherous as the sea. The beginner is taken up for several flights as a passenger. In the initial flight the pilot will perform the most daring manoeuvres and precipitate turns, watching his passenger closely the whole time for any signs of nervousness or fear. It is a most trying ordeal that first trip up aloft, and the bravest hearts have been known to quail.

FIRST FLIGHT ORDEALS

Recently there was a case at a large school of a major of marines, concerning whose courage there could be not the slightest doubt, and who possessed, among other decorations, the much coveted D. S. O. After a first trip above, the major remained in his seat of the landed aeroplane for fully a quarter of an hour, ashen of countenance, and too terrified to speak. It was not cowardice, but simply that he was temperamentally unsuited. At length, when he had composed himself sufficiently to clamber out, he vowed that never again would he go up in an aeroplane.

Following the first flights there are numerous trips in dual-control machines, that is to say, with the ordinary pilot's control-stick and steering-bar duplicated, and both couples working under the same control. Thus, gradually, the "quirk" becomes used to the handling of the craft and accustomed to the sudden drop in an air bank, or to an outward slip in a gust of wind, until eventually, without his knowledge, the instructor allows him to fly the machine himself.

Sufficient progress made, he is allowed to make flights alone, and when he has learnt to bank left and right, and land the machine in a safe and seemly manner, permission is given him to attempt the Royal Aero Club's certificate; for which an altitude flight, a distance flight, and a landing on a given spot are the only tests that are necessary. This, let it be said, is but the starting-point of the flying education. Flying fast machines, wireless operating, machine-gun firing, bomb-dropping, navigation and map-reading are still to be mastered. Only one who has been in the air and seen that queer panorama of jumbled green, grey and blue, stretching away for miles on either hand behind him, can appreciate the difficulties of an air pilot endeavouring to make a true course from a mist-bound earth; or when one's hands are frozen to the bone, and the ice-cold wind whistles by one's ears, the extreme difficulty of manoeuvring the control-stick and working the machine-gun at and the same time.

Reconnaissance and Night Flying

This much for daylight flying, but what of the night when sky and earth are alike indistinguishable? Truly night flying is a science unto itself which needs more than the average amount of courage. However, nightwork is given to only the most experienced pilots.

With active service flying again, we enter into a new phase of which reconnaissance work occupies at least eighty *per cent*, of the time. Simply put, reconnaissance means flying over the other fellow's lines to see what he is about, if he is massing troops at a certain point, or digging in new gun emplacements, or if there is any unusual activity on the highways and railways immediately behind his firing line. It is a difficult matter to differentiate between infantry and cavalry on the march; to distinguish a cleverly hidden gun emplacement, or to tell the difference between an ammunition and a supply depot.

Bomb-dropping is a practice that requires the patience of a Job, good judgment, and a calm day—that is, if it is required to attain any degree of accuracy. Last, but not least, there is the matter of aerial combat, which however, covers too wide a field for discussion in this short chapter.

Thus, in the complete air-pilot, we have a blend of gunner, wireless expert, map-reader, amateur detective, and aviator.

(Part 2 contains a series of incidents and adventures taken from the note-book of a British air pilot, stationed somewhere in the north of France, and are given in their original diary form.)

CHAPTER 9

Behind the Firing Line

Somewhere in France,
Friday.

Tucked away in a corner of an unused Flanders roadway, a long straggled line of irregular shaped huts and sheds surrounding a wide open meadow land, several acres in extent, is the aerodrome I have in mind.

On either side are the long gaunt avenues of trees and in rear of them, bare and low-lying arable lands.

No one can claim for it that it is a beauty spot. But it is comfortable, and above all one is able to obtain a bath there.

On the right are the officers' quarters: three long, low, wooden huts. Within, a passage runs along the centre of the hut; and on either side of it are the various cabins, each about six feet square, and providing just sufficient space for a camp-bed, washstand and chair.

A stove is at either end for warming purposes; and one bath is allotted to each hut.

The mess-room is contained in a similar building across the way. The furniture is not such that one would meet with, say at the Ritz or the Savoy; but it serves its purpose. Three plain deal tables, each covered with a spotless cloth. A dozen or so stiff-back wooden chairs, and one solitary easy-chair. The competition for the latter is enormous.

The general atmosphere of the place is cheery to a degree. Every member of the mess is full of good humour, quips and jests. Sub chaffs captain and captain chaffs sub, the while they attack plain wholesome fare with an unstinted vigour.

After dinner in the evening, an impromptu concert is started. One, an obliging musician, renders an excellent violin solo. He is followed by a gentleman of poor voice. The station orchestra, in which the penny tin whistle is the most prominent instrument, plays delightfully and harmoniously with the possible exception of one member in the extreme rear, who, having previously had some breadcrumbs gently deposited down his neck by an admiring colleague, finds some difficulty in reaching the correct notes. It is, of course, the star-turn of the evening.

There are good card-games to be had, when off duty. Also a gramophone and two pianos. The gramophone usually will not work. Ludo is the rage today. Badminton, writing letters home, and visiting the neighbouring town about complete the leisure time. There is, however, really not very much to do in the town, except to sit in the cafes, drink bad coffee, and try to talk French to the girls.

Any number and variety of pets and mascots are there. Cats and kittens, dogs of all breeds. A few hunters, with which some excellent rides across the sand-dunes can be obtained. A goat that wanders around the aerodrome risking life a dozen times daily from aeroplanes getting off and landing. And a parrot with a perfectly wonderful vocabulary of oaths.

Thus far we have been shown only the lighter side of the life. Now we come to the more serious work of flying across the lines. The strain on the nerves is so great that a pilot is only detailed for duty every other day. The work is distributed among the various squadrons and flights. One is responsible for reconnaissance work; a morning and an afternoon patrol along the coast for submarines, or a trip inland to have a look at a new gun emplacement, or to report on a new movement of the enemy's troops. Another, the fighting squadron, is responsible for the bombing raids, for the battle flights, for convoying the reconnaissance machines, and for meeting enemy air attacks.

To the headquarters flight is allotted the photography, and any special and confidential job that may crop up.

Naturally there is the pick of all the machines, equipped with all the latest improvements and inventions.

One peculiarity concerning atmospheric conditions on the other side is, that either the weather is too misty for flying, or on the other hand, it is so remarkably clear, that it is possible to view the land from twice the altitude that it would be under similar circumstances in England. For the first two hours after sunrise there is invariably a heavy ground mist, and very little takes place save when an expedition is setting out for some distant spot, necessitating an early start. The late morning and the late afternoon are the most favoured times for flying purposes.

Almost the whole of the Flanders country is intersected by waterways and canals. This is of extreme value to the air pilot, and aids him greatly in the matter of navigation. Railway systems there are in plenty, mostly following an east or west direction.

The junctions of these railway lines are the nerve centres of the German Army in the field; they control entirely the supplies of reinforcements, ammunition, and supplies to the firing line. It is for this reason that so many of our own air raids have been made on Bruges, Courtrai, Roubaix, Lille, Tournai, and Douai. Each of these towns mentioned contains an important railway junction.

The large majority of the Belgian towns in the enemy country, immediately behind the firing line, have been totally deserted by their inhabitants and the soldiers alike; it is not considered either safe or desirable to remain within the area of a conspicuous landmark, of which the enemy artillery can obtain an exact bearing with the utmost ease. Added to this, frequent allied air-raids, and the accurate firing of the Allied artillery have reduced them to untenable masses of fallen masonry.

A point regarding aerial photography is worthy of note; if the surface of the earth has been disturbed in any way within two days previous to the photo being taken, that is, disturbed by the explosion of a shell, or a new path across a field made by the tramp of many feet, such disturbance will always show up prominently on the camera negative.

CHAPTER 10

The First Trip Across the Line

Somewhere in France
Monday.

A most important entry in my little diary, this, the day of my first trip across the "lines."

And here in the privacy of my thoughts and of my pen let it be said that at first I was troubled with qualms of fear—qualms that I had experienced in the previous life after a stormy Channel crossing, or prior to a visit to my dentist.

As I stood there on the dreary, wind-swept aerodrome in the chilly rays of the early morning sun, forebodings filled my mind. Visions of an awful death in mid-air, and a yet more awful vision of a downward rush of thousands of feet to the ground below. Comforting myself with the reflections that, after all, out of the large number of machines that must daily cross the lines the proportion of those reported missing was extremely small, I was roused from my pessimistic thoughts by the voice of the pilot, who was already in his seat enjoying the luxury of the last few puffs at his "gasper" cigarette) before testing the engine.

He invited me cordially to "hop in," and once in to strap myself in securely. With his calm matter-of-fact air, which, incidentally cheered me up considerably, one would have thought that we were about to start for a motor run through Piccadilly and the Park rather than, as he so picturesquely styled it, "to play the part of a clay pigeon atop of a firework show."

Three heavy-eyed mechanics now appeared upon the scene, and, after having been slanged roundly for their late arrival by our cheery Jehu, the engine was started with an alarming whirr. A few preliminaries and she got well away.

For a few moments we circled round the neighbourhood of the aerodrome, to gain height. Then in the first contact with the icy-cold morning breeze I felt thankful that I had taken the sound advice of clothing myself well. I must have looked for all the world like an Eskimo or an Arctic explorer in my wool-lined leather coat and overall trousers, a knitted Balaclava hat or helmet, and over that again a skull-cap, the whole tied down tightly beneath my chin. A huge woollen muffler round my neck and a pair of unsightly goggles completed the picture. I had treated my hands and face with a generous dose of Vaseline, which I had been assured would keep out the cold, and which advice I now gratefully acknowledge to be correct.

As we mount higher my perspective extends and out of the grey mists and the dark shadows land and sea begin to assume their natural form and colour. On the former there are now signs of movement; along the roads crawl the ant-like procession of ammunition columns back from their nightly trip to the firing lines. A steaming "Puffing Billy" slowly drags along on a limber, a "grandmother" (naval 15-inch gun) blocking up the whole roadway, which must cause considerable annoyance to the long string of cars and motorbike despatch riders held up in the rear.

On the roadside, by a wood, a company of infantry are falling in for early parade; they look up at us in a half interested sort of way. Some wave their hats and rifles at us. I wave my hand in reply, but know they cannot see us. We keep on climbing steadily. Out at sea are two French torpedo-boats making up the coast towards ——, and a few small trawlers sailing off in the direction of England. Happy thought!

Every moment we are getting nearer to the dreaded area. In the far distance I can see the red flashes of the rifles, the smoke clouds of the heavy guns, and the long grey lines of winding trenches. I look at my map, to discover that we are passing over a junction of two main roads, one of which is crossed by a railway, while beneath the other runs a narrow stream. It is ——.

Five miles to the firing line. With my glasses I can already pick out several of our own field-artillery emplacements, and a moving up of reinforcements from the rear—I would surmise about two battalions of infantry. I time the observation on my report sheet; also I discover from my wrist compass—my most prized and valued possession—that we are going too much to the north-west and tell the pilot so by means of a written message.

Course changed! What are Headquarters' orders for the flight? A reconnaissance over ——, I puzzle out as well as my now fevered brain will allow me, whether reconnaissance will be tactical or strategical, and again whether "line" or "area." For the benefit of those who may perhaps read my diary I will here endeavour to explain the fine points which divide the two.

The former reconnaissance necessitates flying and observing along a line between two given points on the map, these points having already been marked in before leaving the ground. Area reconnaissance, on the other hand, comprises observation of a whole area or district. To do this successfully it is necessary to fly backward and forward several times, thus adding greater risk to the adventure, and taking a great deal longer time to accomplish. Hence they are not undertaken very far away from our own lines, and then only if particular information is required.

Thus far the weather had rendered the trip ideal. But it would be an entirely different matter, I surmised, when we came within reach of the enemy anti-aircraft guns. Already they were getting uncomfortably near. Should we have an easy passage across or should we have to climb up for our lives above the bursting "Archies?"

We were not left long in doubt. Their men must have been up particularly early that morning, for the very first shot came within an ace of blighting two young and promising careers. There was a loud report on the ground below, the familiar "*sing*" of an approaching shell, which at first interests one, but which in the course of time one gets to dread. Then it seemed for the moment that the whole machine had been blown to atoms. But no! We started to climb hurriedly.

"High explosive" the pilot bawled in my ear. "Going up higher."

For the next three minutes my feelings were the reverse of pleasant, and I fervently hoped that other observers did not suffer in the same way. Shells burst above, below, to the right, to the left, and all round us; but never near enough to do us any serious harm, though the bullets of one shrapnel shell certainly did rattle against the wings, piercing them with minute holes in several places, and I felt very thankful for the uncomfortable sandbag on which I sat, which protected me from bursting shells beneath.

As we climbed to a higher altitude the Huns ceased their attentions, and we very soon arrived over the scene of our "line." My bad attack of "cold feet" now having passed over, I set myself to think seriously upon the precepts drummed into my thick head by the instructor at the training school. "The observer" he was wont to say, "should always try to keep in touch with the military situation, and particularly in the encounter battle, and discover the disposition of our own troops."

One point I could and did satisfy myself upon—this was no encounter battle. So I ignored our own forces and kept my attention fixed upon ———. Nothing extraordinary met my eye. I saw a camp here and there, and turned my glasses upon them and discovered that they were composed of huts. Hurriedly I counted them, and noted the number in my report, together with the altitude, 12,000 ft. Again the solemn advice of my worthy instructor passed through my brain: "The eyes must constantly turn to each likely spot, and each spot must be examined carefully with the glasses if it offers anything useful for the observer's report."

I examined each likely spot, and discovered to my delight a broad grass meadow across which ran several pathways of very recent construction. Footpaths, I argued to myself (and I may possibly have been wrong) are not made across fields for the mere pleasure of constructing them. There is more in this than meets the eye. I signalled to my companion and he quickly grasped the situation, and in long sweeping circles, brought her down some 2000 ft. The lower we came the more distinctly I could make out that some sort of emplacement was being built up—the new emplacement for a 17-inch howitzer. I noted the same.

An excellent morning's work. We turn to go home. But the enemy has not appreciated our attentions and most unthoughtfully turns his guns upon us.

Then the fun begins. It was bad enough crossing the lines, but child's-play when compared with this; and besides we are two thousand lower. A perfect inferno of "Archies." We bank first to one side then to the other; put her nose down for a moment or so, then climb for all we are worth.

But it is no good. We are hit!

Down goes her nose, down and down. The air whistles past our ears. The earth rushes up to meet us. The discs of the machine-gun topple overboard, so steep has the angle become. ——— must have been hit. Yes! there he is, all huddled up over the joy-stick (control-stick). I give up all hope, when suddenly, the machine, starts to right herself. I look around, and find that the rush through the air must have brought him to. He is manfully straining every nerve to get her out of the nose-dive. By a superhuman effort he succeeds. We manage to get across the lines unnoticed save by a few infantrymen, who fire futilely at us, and land a bare hundred yards the other side of our own trenches. —— makes a beautiful landing, pulls her up dead, and promptly faints in his seat. My first trip!

CHAPTER 11

Some Anecdotes

Somewhere In Belgium,
Thursday.

"The life of an airman is one of intense idleness interrupted by moments of violent fear." This remark, originating as it does from a youthful member of the Senior Service describes, more aptly than any other yet penned, the life of the airman under active service conditions. Sometimes there will come a spell of fine weather, and he is kept going hard at it from sunrise to sunset. At other times when the weather is too bad for flying, he has nought else to do but sit round the mess-fire and tell stories.

The memory of those wet days! Men of all sorts and conditions exchanging personal experiences: anecdotes of hair-raising escapes from bursting shrapnel shells, thrilling fights with Air Huns, miraculous evolutions in mid-air, and a thousand and one other subjects dear to the heart of the airman. I will here endeavour to relate several of the best stories that have so far come my way, but it is impossible to tell more than five *per cent*, of them, for their name is Legion.

The first story concerns a well-known aerodrome somewhere in Flanders. The pilots of the station, when the weather was too bad for flying, filled up their spare time by playing football; until one day a wag amongst them suggested that a ball should be blown up as tight as possible; taken up in an aeroplane and dropped on the German lines. This suggestion was duly carried out and the first fine day the ball was put aboard a machine going up the Belgian Coast for a reconnaissance trip. Arrived over the town that had been decided upon, it was dropped overboard, with quite accurate aim into the market square. Seeing this dark awesome object falling through the air and taking it for a bomb the Germans took to their heels. Landing on the cobbled pave, it must have bounced nearly twenty feet into the air, then gradually lower and lower, until at last it rolled into a ditch. Then and only then did the Germans reappear, one fat soldier going over to it and giving it a vicious kick.

An instance of air *camaraderie* was that of the Bosche who brought Pégoud down after a fight in mid-air. Hearing that he had been killed, and where he was to be buried, he came over and dropped a wreath on the scene of his burial ground—a pretty compliment that was greatly appreciated. The story concerning Captain M—— is the most striking of the war. Poor fellow, he has since been killed. It happened one very misty morning. M—— was on a reconnaissance trip. His engine failed and had to come down a good ten miles behind their lines. However, he landed safely, and had just burnt his machine, when he saw three dark figures coming up out of the fog, and taking discretion to be the better part of valour he fled, and hid himself in a ditch hard by. He was there for the whole of a day and a night, and it has since been ascertained that there were close on five thousand Bosches searching for him the whole time.

When he found the coast was clear, he crept out of the ditch, and marched off boldly down the road until he met a friendly Belgian peasant; from this chap he wheedled an old suit of clothes, and, thus attired, walked on nearly to Lille. Here he acted somewhat foolishly. He boarded a tramcar bound for the city, not knowing where to ask to be put down. The car was full of Prussian officers. The man came for his fare; and for a moment he was nonplussed. Then he had a brainwave. Remembering that every town in Belgium possesses a glorified market square, he demanded *à la grande place, s'il vous plait*, and pulled out a handful of silver coins to pay the man. Such a thing as a silver coin had not been seen in Lille for months, ever since the Germans had captured it in fact. Fortunately the Prussians were too much occupied in their own conversation to take any notice of *ein schweinhund* of a Belgian peasant.

Arrived in the city, luck again favoured him, and he obtained shelter in a garret for three weeks. Then the police grew suspicious, and late one night he was forced to clear out hurriedly. After leaving the city he had a terrible time. He tramped right across Belgium, always at night, and every moment in fear of his life, feeding on anything he could find, crusts and offal thrown to the pigs, and stale bread thrown away by the German soldiers. Footsore, weary, hungry and exhausted he at last arrived at the Dutch frontier. Here occurred another agonising wait.

Again for a day and a night he lay hidden in a ditch, until late that evening the sentry paused on his beat to light his pipe. This was his opportunity. It was a moonlight night. He dashed across the intervening space. The sentry fired three shots and missed each time. He got across Holland, to a seaport town, stowed himself aboard a fishing smack, got to England and reported himself to the astonished officials at the War Office, This reminds me of a story told by a certain famous airman, a little man with a great heart, on whose breast there are the flaring crimson of the French Legion d'Honneur and the crimson and blue of the Distinguished Service Order. I will give you his own words.

"I went over the lines with X—— for an observer. He'd never been over the lines before and I must confess that I felt a wee bit shaky as to how he would take it. Luckily we got across without a single shot being fired at us and then we met a Taube, coming right down wind at about ninety miles, and at about our own level. I looked at X——, who for a time, was too busy watching the other chap coming up to notice me, but finally he turned and smiled, and I knew he was all right. 'Got the Lewis-gun ready?' I bawled into his ear. He nodded, and then we cleared the decks for action, so to speak. He put a fresh tray of ammunition on the gun, and got two other trays ready by the side of him, while I had a look at the bombs and grenades, and put the joy-stick about a bit just to see that she was all right.

The other chap still kept on, and was only about a hundred yards off when X—— opened fire, *zipp-zipp-zipp-zipp*, seventy-eight of the little beggars slick into the middle of him. Gave him hell, I can tell you; at all events he didn't stick it long. Down went the nose of his machine, and he was very soon about a thousand feet beneath us. I loosed off all my bombs, quick as I could, missed every time, had a shot with a grenade and missed again. I must confess I felt a wee bit flurried that morning and then X—— began. Never laughed so much in all my life. He laid his hands on everything, his hat and his glasses—government glasses, by the way—and his revolver and spare cartridges. Thank God! There was nothing of mine in the front there."

Not nearly so pleasant, however, was the experience of a certain seaplane pilot, who when flying across the Channel from Belgium to England was forced by engine trouble to come down on to the sea, in the midst of our own mine-fields, very far removed from the track of all shipping. Here he remained for eleven and half hours, until sighted by a torpedo-boat, which though unable to reach him herself, was able to give warning ashore, so that a small motor-boat succeeded in finding a way through the mines, rescued the pilot, but was forced to abandon the machine.

Another story concerning Pégoud. The Germans brought Pégoud down, when flying one of the new French machines, that are supposed to have so many wonderful new improvements aboard, and that they're so secretive about. He didn't have time to burn it—and the Huns were very keen on learning how the thing flew. So they tackled Pégoud on the subject. He said he was perfectly willing to give them an exhibition himself, but they didn't care for the idea. "Yes, and when you get up there you'll fly away back to your own lines again." "Very well," he said; "send two of your men up with revolvers and let them sit one on each side of me, so that I won't be able to get away." To this they afterwards agreed, and the first fine morning Pégoud, with the two men sitting on each side of the fuselage, goes up about 10,000 feet. Then one of the Huns began to get impatient. Said he: "I think we'd better go down now." "That's all right," Pégoud answered, "you're going." And with that he put his joy-stick down. She went over a good clean loop, and the Bosches went down quicker than they bargained for.

CHAPTER 12

Sport Extraordinary

Somewhere in the North of France,
Monday.

There is an undoubted fascination in being about at sunrise on a clear, fine morning. And especially so when up in the air.

Our day was of this variety. A day when a man's heart yearns for a moor, a dog, and a gun. For moor we had the long, flat, dreary sandhill and marshes of the Belgian coast; a dog was not needed, and in fact would have been in the way.

And our gun was not of a type particularly well-known or approved of in sporting circles—a "Lewis" machine-gun, fitted above with a tray of forty-seven cartridges.

Our quest was "wild ducks," an idea as novel as it was entertaining, originating with the *padre* of the station—a cheery individual, who divided his attention between writing insufferably bad verse, and collecting mess-subscriptions from irritated members.

The sun rose over the sea, lighting the blue surface with a thousand scintillating rays. The tents of the camps thousands of feet below began to show up against the grey of the earth, and the red flashes of the rifle volleys combined with the white cloud and roar of the belching heavy-gun to complete our picture of the waking world.

But we had not much time to pay attention to these matters, for our minds and eyes were concentrated on the one subject.

From what direction would they first appear? Would they come up to us, or would we have to put "her" down to them? The sun was well up in the sky, and signs of life and movement were beginning to make themselves manifest "down there," before several tiny black specks appeared on the horizon coming up from the ground behind the marshes at Nieuport.

We brought the aeroplane round, to get the birds between the sun and ourselves, and with the wind at their backs, so as not to be aware of our approach. However, they turned off seawards, and again we had to change our course, until they seemed to be at too great a distance for us ever to get them within gun range. The noise of the racing engine must have reached them on this new tack, for we were now only half-head on to the wind; but of this they took not the slightest notice, keeping on their way a regular and well-ordered flock.

As a matter of fact this could be explained by the reason that birds in that neighbourhood must have become so entirely used to the whirr of a passing aeroplane, for as many as a score passed over this same district every fine day.

We now changed our tactics, and brought her round with the sun at our backs, casting a shadow across the path of the moving flock, and a small dull replica which moved in an alarming and amazing manner across field and hedge, house and farm, beneath.

At last we were getting up with them, and to signalise the happy event the *padre* let off a dozen rounds, which went very far wide of the mark, and only served to divide the flock into two portions, the larger of which continued in a seaward direction.

These we determined to follow, and coming down to 500 feet, opened the engine "full out" to close on 100 miles an hour.

Never before had one realised the wonderful speed which these birds can keep up when on the wing. For with all our great speed we were yet far behind, and every moment drawing nearer to the sea, across which at this extremely low altitude we dare not venture.

Thus it seemed as if we should have to return, defeated and discomfited, to a scoffing, chaffing audience on the aerodrome, still visible some five miles to the south-east.

However, immediately before reaching the seashore our quarry turned again, and this time along the coast. Then, banking her over to the new direction, we found ourselves "down-wind" with an additional speed at the back of us of 15 m. p. h., which soon began to tell. The *padre* began to get unduly excited, and succeeded in giving a not unmusical series of "*zimms*" on the gun; the cartridges falling spent and useless on to the sand-dunes; there were no casualties. Undaunted, we kept on, taking care this time to get nearer up. The enemy were beginning to tire by this time, so putting in a fresh tray of ammunition, our courageous marksman let fly, with excellent results, three of the rearguard speeding headlong down to the earth. The pangs of a not unnatural hunger now beginning to make themselves evident, and finding ourselves some thirty miles from home, we turned her head for home and there eventually arrived, happy and hungry, after having set a new fashion in sporting and aviation circles, and discovered a new form of amusement and speculation for the *blasé* ones, who had deserted their card-tables and cheap French novelettes to welcome us on our return.

CHAPTER 13

A Balloon Trip By Night

Imagine a great bare meadow-land, lonely, windswept, and dark with inky blackness, out of which there plunges an occasional hurrying figure, that misses one by inches and passes on with a muttered oath. In the background, tall and sinister, two large gasometers. In the centre of the field a wide tarpaulin laid along the ground, and edged by a circle of sand-bags, from the midst of which there rises a great round shape, like a mammoth tomato.

It is the balloon not yet fully inflated, fed by two curling rubber tubes, that disappear in the direction of the gasworks. We are waiting, waiting patiently until she fills. Blackened, distorted shapes, that stand around in eerie circle, and at the sudden gruff command of a hoarse voice that booms ever and anon out of the voids of darkness, seize each a heavy sandbag and slowly and clumsily lower it mesh by mesh in the netting that covers the balloon.

At last she is filled. The car is attached below, as rapidly and securely as the faint and flickering light of a stable lamp will allow of. The crew tumble in, one on top of another. She is let up only to be pulled down again with a nerve-racking bump. The gruff voice decides that she is now ready to get off; there is a slight slackening of ropes, an almost imperceptible lift, the figures on the ground recede rapidly, grotesque shadows in the darkness, and the lights begin to disappear one by one.

We rise to a ticklish situation; there are tall trees, factory-chimneys, and protruding roofs all waiting calm and invisible in the night, to be crashed into and collided with. But all these obstacles we may miss if we have only sufficient preparatory lift. We are all silent and cowed, trying to make out each other's faces. There is a sudden tearing sound. The craft lurches like a drunken man and we are thrown a struggling breathless mass into a corner. But the suspense is only momentary. By a miracle of grace, she frees herself from the branches of a tree, and soars rapidly heavenwards.

Eagerly we watch the glimmering, winding streak of grey that is the river, and our only visible landmark; apparently we are making off in a north and west direction. Once out of the shelter of the houses and the trees, the breeze is stiffish: in fact, considerably more so than was expected.

What is this sensation like? Dark to the left of us, dark to the right of us, dark on top of us, and darker below us; in a frail uncontrollable craft, that drifts aimlessly and helplessly before every varying wind of the heavens. Unlike the aeroplane the passage is easy and pleasant, free from noise and we know we are flying. North and west, but the first change of the wind, and we will be bowling along merrily in quite another direction.

It is quiet, intensely quiet, no motion of any kind to be felt. But where are we? Occasionally we discover a small patch of light that may be a village, again a larger patch, evidently of a town. We watch the altimeter with as much loving care, as a mother would her child, for it is our sole deliverer from destruction. How it varies: now it is 8000 feet, now 2500. If possible, we try to keep above the latter level. The surface of the country is unfortunately not too level, and as the altimeter registers height above sea and not land level, allowance must be made. Ballast is ready to hand for emergency uses.

At last the depressing silence is broken; one youth, wiser than his years, has remembered to provide himself with food. It is handed round, and over beef sandwiches we get communicative. It gives us fresh life and inspires one of the party with a humorous turn of mind, to recite with great vividness and vivacity all the alarming accidents that have befallen night-balloonists, concluding with an impious hope, "that we likewise may have some fun."

We get it!

Happily, as we are wallowing in the throes of this most dismal expectancy, the conversation is turned by an eager and heated discussion between two younger members of the party, as to the merits and demerits of their respective musical-comedy idols (female). The argument grows in intensity. But we have neglected to watch the altimeter. Out of the inky darkness below there rushes a volcano of spark and flame. It is a railway-train speeding on through the night. Sheepishly we discover that we are only 800 feet, and wonder unpleasantly what might have been.

On and on through the night. Now we are getting tired; there are suggestions that we should land, but they are overruled. Coming down again to 800 feet, we catch sight of a wide glimmering sheet of water. Maps are seized in a hasty impulse to guess our whereabouts. The argument grows heated, for similar stretches of water there are, alike in Essex, Kent, Surrey, Middlesex and Berkshire: in fact, in every one of the Home Counties, and for the matter of that in the Midlands, and likewise in every county in England, Scotland, Ireland and Wales.

The argument abates, our eyes grow weary and more weary. It seems a lifetime since we last saw the pleasant and undulating lines of the earth. One or two heads are already nodding, when there is a sudden shout of "the dawn." Instantly all are wide awake. There sure enough, are the first few streaks of grey creeping slowly across the eastern sky; without even that, it would be an obvious matter, by rea-son of that intense cold, which, in the air, always precedes the hour of daybreak and freezes us to the bone.

It would be an inadequate expression to say that dawn in the air is beautiful. It is more than beautiful, it is wonderful. It is more than wonderful, it is unusual; a view only to be enjoyed by the minority, and that of the smallest. Gradually earth and sky begin to dissemble. In tint the picture is white, black, grey, blue, crimson, golden, purple, green and every other colour—now like a painter's canvas smudged with regular irregularities, edged with red and grey, now an animated panorama stirring with resuscitated life. The sun rises, a ball of flame above the horizon, lighting up the rotund shape of the balloon with an unearthly hue.

We say nothing, but look and marvel; a word would be out of place in this sacred and awesome stillness, Suddenly we are roused by a cry, more, much more, alarming than the last.

The sea! We are almost on top of it. In shimmering, level surface it stretches on into obscurity. We are lost. We cannot avoid it, yet less can we land thereon. One of the crew loses his head. He snatches the thin red tape that hangs down from the envelope. There is a tearing, rending sound.

He has ripped the balloon at 2000 feet. Pious prayers and curses intermingle. Down she sinks, with a great hole rent in her side down and down, faster and faster. Over go the bags of ballast, one after another. Now all have been dropped. She slackens speed; but only momentarily. Down she goes again, the upward current of air whistles unpleasantly through the rigging. In a last feverish effort boots are unlaced and hurled overboard, together with coats and every portable object to hand.

Too late. We hit the edge of a cliff; bounce back several feet into the air, then sink down on to the beach below. Another crash, again we are bundled and bounced about in the confined space of the car. The sand gets in our ears and eyes and mouths. The balloon lies along the sand a woe begotten shape, as flat as a pancake. When we eventually sort ourselves out, we find luckily, that there is but one casualty: a broken wrist, sustained by the foolish idiot that ripped! Just retribution!

And to end the adventure, a stolid British policeman, ponderous official-looking note-book in hand, approaches and demands our names and addresses, and asks if we are of British nationality!

CHAPTER 14

The Battle of the Wood

Flanders, Wednesday.

Somewhere in the north of France there is a little wood. It is about half a mile square in area, and stands immediately south of a fine, broad highroad, along which there daily pass large bodies of reinforcements, infantry and cavalry, and convoys bringing up ammunition and supplies. The tall trees offer a welcome shade in the hot weather, and it was the custom for passing troops to halt there for a short time; and just at the spot the roadside was always well littered with broken bottles. Needless to state, it was in German territory.

However, had it not been for that road, and for the fact that on this certain day, when the road had been closed to all traffic, there were certain mysterious movements of ponderous great wagons, suspiciously like ammunition wagons, which halted in the shade of the wood, this story would never have been written.

The day was hot, and the work was heavy, and *mein herr captain* paused for a moment to curse his uncongenial task, and take a long draught from his water bottle, of some liquor that certainly was not water. In the midst thereof he let it fall with a curse of rage and surprise, for there overhead, as if it had suddenly appeared from the clouds, was the form of a British aeroplane. "*Himmel,*" he exclaimed, "all our trouble wasted, they have our hiding spot discovered, and to-morrow morning they bomb us—*ach!*"

The worthy gentleman was not far out in his deduction, for the lynx-eye of the observer in the aeroplane had carefully noted the exact geographical position of that new ammunition park, before the machine sped off homewards. But he was wrong to a certain extent; our Flying Corps are no fools, and they realised that Mr. Bosche would soon expect a return visit, and would be fully prepared therefore. This

64

course was, therefore, useless to them; it was essential that that ammunition park must be destroyed, but in a manner and at a time the Germans least expected, and this is how it was accomplished.

Towards evening a light scouting machine sped swiftly away from a certain British aerodrome, only a few miles behind the firing lines. No untoward incident that, but it was particularly conspicuous from the fact that the entire aerodrome had turned out to wish the trip God-speed, to wish the pilot, a young second lieutenant of the Canadian Infantry, the best of luck, and to cram the fuselage of the machine with spare ammunition, until she could barely "stagger" off the ground. The objective was the ammunition park already mentioned. With long, sweeping circles the scout soon cleared the area of the firing lines, and arrived over the wood.

Still nothing happened, the whole countryside was remarkably quiet for a battle area. No anti-aircraft guns fired, no enemy aircraft came humming round. Lower came the pilot to investigate. Still nothing happened; he, on his part, now began to feel genuinely alarmed, unless of course that confounded observer had been "seeing" things, a not unknown failing with aeroplane observers.

Meanwhile in the midst of the wood, the corpulent captain watched the small speck carefully with his glasses, then rubbed his fat hands with glee and expectation. The fool Englishman was falling beautifully into his little trap. Involuntarily he glanced over his shoulder, and there in a large clearing behind the wood, were ten great German battle-planes, all ready to go up at a moment's notice and with pilots and observers standing by.

By this time the British machine had come considerably lower, and was well behind the wood, and into the German country. The captain gave a sharp, guttural order. Immediately the noise of ten great propellers smote the still air, and the squadron rose swiftly from the wood like a covey of wild ducks. The hated Englishman was hopelessly trapped.

And what of our man? Turning leisurely to make a last reconnaissance of the wood, he found ten great German battle-planes between himself and the lines. He cursed profusely at his own crass stupidity. He had been warned, and he had thought fit to ignore the warning, and this was the result. Anyway he would make a good fight for it. He fingered his machine-gun cautiously. Yes, everything was ready to hand. He set his teeth, opened his engine "full out" and began to climb rapidly.

The Germans also climbed, and within a very short space of time he found himself hemmed in on all sides, with lead flying at him from all points, and at all angles. Anyhow, he determined to have a good run for his life, and singling out two Germans immediately beneath him, he dived rapidly. As he did so, he was hit by shrapnel; for a short space of time he was unconscious, then again regaining control of his machine, began to use his machine-gun to good effect.

First one German he drove to the ground, then another, and then a third. His blood was up now, and he turned round for further victims, but Huns had had sufficient for one day, and were scuttling off to peace and safety. He turned homewards, and his wound was becoming agonising, as a bombing squadron of our own machines passed by.

Very soon there arose from the wood violent explosions and blinding sheets of flame, and by the time the British bombing squadron had finished its full design, all that remained of the fat captain's ammunition park were a few broken and shattered wagons, and a heap of dead and dying men.

CHAPTER 15

A Tight Corner

Somewhere in France,
Friday.

The other day, yesterday afternoon to be exact, a most exciting adventure befell me. I was detailed to take part in a bombing raid at ———. We had not proceeded far beyond our own lines, after the customary bombardment of anti-aircraft shells, when suddenly the machine immediately in front of us rocked violently, and began to dive towards the earth. "B———'s been hit," my observer bawled into my ear. I continued to watch the machine in its headlong descent. Alas, it was only too true! There was no possible escape: after diving steeply six hundred feet, the machine had begun to spin, and was now whirling round and round like a humming-top, and hardly a minute after, had crashed into the midst of a wood, from which there immediately came up a cloud of grey smoke and leaping tongue of flame.

We had started out four strong; our mission being to raid M———, a large German military centre, containing a staff headquarters, an ammunition park, and a large aerodrome. And now our machine was the sole survivor, two having been shot down when crossing the lines. Alone and single-handed, in a notoriously dangerous portion of the enemy's lines, every moment we were liable to be fired at from all quarters, and attacked by enemy aircraft.

I looked searchingly at my observer; it was his first trip across the lines, and I had to admit to myself that never before, in my six months of flying at the front, had I been in such a deucedly uncomfortable position. How would he take it? I hesitated. Should we turn back to safety, or should we continue on our way to what was almost certain death? I glanced at his face, it was stern and set, with the deliberation of the man who is willing to risk everything. With his left hand he patted and fondled the deadly machine-gun. I determined to go on.

Then they opened fire on us again. Apparently for the last few minutes they had all deserted their guns and had been busy gaping at the remains of poor B——'s machine; but now, flushed with their recent success, they commenced to fire with demoniacal fury. Shots burst behind, before, above, below: one minute immediately over the nose, the next immediately beneath the tail of the machine. To avoid them we climbed, and dived, and banked in all directions, until her old ribs began to groan and creak from sheer exertion, and she threatened every moment to fly asunder in mid-air.

At last we got clear of them, and sighted our objective, just as the sun broke through the clouds, and revealed to us a stretch of low, flat-lying country, dotted here and there with villages and camps and ammunition bases. M—— showed up easily, it was a moderate-sized town of ant-like pigmy dwellings, little white and grey patches in the brilliant sunlight. A small winding river skirted the town, looking for all the world against the dark background like the vein in a man's arm. North and south ran the gleaming, glinting railway lines, and a large road led up from the town to the firing line. This road was now converged with traffic of all descriptions. We dropped a bomb, but it was very wide of the mark, and it served to draw the enemy's fire, which again broke out all round us with renewed fury. M—— was better supplied with anti-aircraft guns than any other position on the German front.

Higher and yet higher we climbed until we were well above the clouds, and the earth was almost hidden from our sight. By this simple and expedient *ruse de guerre* we might be able to get over the city before the gunners were aware of our existence. But alas for our well laid plans! We had not gone far when we encountered a great double-engined Albatross, and there, with the white billowy clouds stretching like waves of a gigantic sea in all directions, we fought our battle of life and death. Fritz opened the encounter by sweeping down upon us at top speed, pouring out a steady stream of lead from the machine-gun in the nose of his machine. To avoid this we climbed rapidly, and he flashed by, beneath us, at an alarming rate. We attempted to bomb him, but it was futile, and the bomb fell downwards to the earth below.

We turned as soon as were able, and waited for the enemy to recommence the attack. He was all out now, and putting on top speed bore down upon us with the speed of an express train. Nearer and yet nearer he drew. Thankfully I noticed that we were both at the same altitude. When yet about a quarter of a mile distant, his observer opened fire, the bullets flying all around us in a leaden stream, and still we did not reply. I looked at my observer. He was bending over his gun, fumbling about with some portion of the mechanism.

There was no need to ask what was the matter. Alas! I knew too well. The gun had jammed. Now followed a ticklish time for both of us, for without the gun we were completely unarmed, and Fritz was drawing nearer every second. Already I could hear and feel his bullets singing past my head, occasionally chipping portions of the machine. Now he was right level with us. What were we to do? To remain in that same position would mean certain death. If we climbed, he would climb faster, and would almost immediately be up with us again. There was only one thing to be done—the unexpected! So putting her nose-down, we dived towards the earth like a stone, and had gone over a thousand feet before I could get her level again.

This manoeuvre so upset the calculations of the enemy, that he was now about three quarters of a mile distant. This gave us precious time to prepare again for the attack. The observer was still working feverishly away, when we commenced to climb. Fritz had already turned and was coming down to meet us; but we had the advantage this time of having the wind behind our backs. If only that infernal gun were ready! Up we climbed, and down came Fritz; all the faster because he knew we were comparatively unarmed. Now we were under half a mile distant, now only a quarter, and now he had commenced to fire. Would we never reply? At last! *Brrr! Brrr! Brrr!* yapped the gun in our bows.

Fritz was so startled at this unexpected development that for a moment he paused in his firing. This was our opportunity; taking steady aim J—— put the whole drum of 47 cartridges into his back in three bursts. He staggered and reeled, he was hit; I felt I wanted to cry out for sheer joy, but my throat was parched and dry. Oh! the reaction after that dreadful ten minutes. But although we had hit him, Fritz was yet by no means out of running, that is if he elected to remain and fight it out, which I doubted extremely; for the Hun is ever super-courageous when he has an unarmed and helpless foe to deal with. So throttling her down I watched him anxiously. Turning to the left he started off at top speed in the direction of his own base. This I had expected, and off we started in his trail with only another half-hour's petrol in our tanks.

On and on he flew, over wood and town, and we were close in the rear, both flying at top speed. Every moment he was getting lower. I knew only too well what that meant. He was trying to lead us into a trap, where we would make a set target for a ring of his anti-aircraft guns. We must never let this happen or we should be finished for a certainty. If we could only catch up with him; but it was in vain we wished, for he was yet a quarter of a mile ahead, when, as usual, the unexpected happened. He had engine trouble. Within five minutes we were almost on top of him. He commenced to sink like a stone.

Now was our opportunity, an opportunity which our observer was not slow to take advantage of. Right into the middle of his back flew the steady stream of bullets. Again he reeled, and this time there was that peculiar fluttering of the wings, which tells only too plainly that an aeroplane is "out of control." Like poor B—— he commenced to whirl round like a humming-top, then with one long last plunge he had crashed into one of his own encampments, and all was over.

We were left to reach our own lines with twenty minutes' petrol remaining, and under a violent bombardment of the enemy "Archies."

★★★★★★

Again an interesting personal account, told in the words of the pilot participating in a Zepp Strafe:—

The orderly from the telephone room brought the news. Zeppelins had been sighted at —— and were proceeding in a northerly direction. This meant that they would be overhead at any moment.

A few sharp orders and the station began to throb with life.

Mechanics hurried hither and thither, some to the sheds to get out the machine, others to fetch the bombs and a Very's pistol from the armoury; yet others to lay out the light flares across the aerodrome in order that upon our return we might perchance be able to define the right landing ground.

Compasses, electric light torches and maps were dragged hurriedly from their hiding-places in lockers. A general bearing was taken of the enemy's course, and we ran out on to the aerodrome, where a searchlight had already begun work, sending long, scintillating beams of light across the dark night sky, turning and twisting, first in one quarter, then in another, covering the heavens in the twinkling of an eye, but never disclosing the true object of its search.

At last there is a shout from one of the men by the light. He had discovered the whereabouts of the Zeppelin. Yes! there she is! A long, grey cigar-shaped object far up in the clouds.

We hurried across to the machine, and while I examined the bombs in the bomb-rack beneath the fuselage (body), and attended to the fitting-in of the Lewis-gun, the pilot tested the engine. And before five minutes had elapsed since the first alarm we were off the ground.

Who can well and truly describe the sensations of night flying? Suddenly one is hurled from the ground into an unbounded space of darkness at the rate of fifty miles an hour. It is like jumping off a cliff on a dark night and plunging on and on, one knows not where. It is impossible to see beyond one's nose, and the only thing that seems real and natural is the incessantly loud hum of the engine. It is a by no means pleasant task.

Leaving the ground we miss a roof-top by inches, and, feeling considerably shaken, climb rapidly. At first it is dark, pitch dark. We see nothing, we know not where we are. One would lose one's reason were it not for the hum of the racing engine.

At last there breaks through the long shadows of darkness, beneath us, a long, narrow, winding ribbon of shimmering grey. The young moon has broken through the clouds and the reflection of its light upon the water gives us the position of the river. On either side or moving slowly along the surface are small pin-pricks of coloured lights; I switch on my electric light in front of the observer's seat, glance at the altimeter, and discover that we are already 500 feet up.

The glare of that light, feeble though it be when contrasted with the black darkness of the atmosphere around, has got into my eyes, and for a moment or two I can distinguish absolutely nothing. Then lights begin to make themselves visible.

The street lamps can easily be distinguished; as being darkened at the top the light is concentrated downwards in a circle on to the pavement beneath, which serves the purpose of reflecting it heavenwards and upwards. The main streets can be picked out by the two parallel lines of coloured lights; the windows of shops, the lights of which have been covered with red and green shades.

I have another look at the altimeter. Only a thousand, but still climbing steadily. Into a dark bare patch of land far below there comes rushing a flaring, glaring gleam of light, followed by a string of smaller lights. I puzzle out what this strange apparition may be. It is a railway train.

As we mount yet higher we begin to lose all our bearings, and all sight of the earth beneath. A much more beautiful earth when compared with the dull, prosaic everyday affair, looking for all the world like a huge garden decorated with a myriad of multi-coloured lights. It is difficult to realise that those few, straggling, irregular rows of lamps encompass seven million living souls; that there far below us sleepily blinking and twinkling is the greatest city of the world.

The altimeter registers 5000 ft. Getting nearer to the Zepp altitude, yet no sign! The anxiety of waiting and suspense is becoming insufferable. Nothing but the incessant throb of the engine. But I have spoken too soon! Out of the darkness and blackness there rushes past, with the speed of an express train, a black unholy shape.

Suddenly there is the most violent cannonade; a sure sign that the anti-aircraft gunners have spotted their quarry. Searchlights from all directions are in a second of time concentrated upon ourselves, while they are endeavouring to get the range. This latter, much to the disgust of the pilot, who, blinded by the glare, banks too steeply, just in time saves her from a nose-dive, and consigns all anti-aircraft gunners to a certain well-known locality possessed of a permanent and extremely warm climate.

We are in luck's way, however; for presently the guns are all silenced. The searchlights go out one by one. All becomes quiet and dark, dismally dark. We cruise around for another ten minutes or so, then descend cautiously and gradually. With one eye glued to the altimeter, to make certain of the height, I peer over the side with the other to pick up the first sign of lights or landmarks.

Eight thousand feet! Seven thousand feet! Getting horribly cold! Six thousand! Five thousand! Shall we never get down? Four thousand! Three thousand! it seems like an age. Two thousand! One thousand! Cautiously now or our necks will be broken!

At last we are safe back on Mother Earth again, and very thankfully seek the refuge of our beds!

CHAPTER 16

An Air Fight With a Hun

Somewhere in the North of France,
Saturday.

Today our special delight has been a bombardment from enemy aeroplanes.

They came over about noon and roused the fearful and subdued the proud while we were all at lunch. They circled overhead for about five minutes, dropped a dozen or so bombs, then cleared off hurriedly before our own men had time to get away.

One man here had a most ingenious "funkhole" for aerial bombardment. He utilised a large stone drainpipe for this purpose, and it was his custom when enemy aircraft were reported to be in sight to crawl into this thing, take a book with him, and calmly read until they had taken their departure. He advertised this comic shelter one day as:—

A novel *bijou* residence, completely detached, every convenience, within easy reach of the firing line. Bullets and bombs pass the door every few moments.

Figuratively speaking, our mission was target-registering.

But having previously heard that the "mother" (naval 9.2-inch gun) with which we were to have worked was incapacitated, and the afternoon being fine and sunny, we determined to seek adventure further afield, and turning her nose in a south-easterly direction kept straight on.

Am making for Dixmude[1] to see if we can raise a Hun or two.

---

1. *Dixmude: French Marines in the Great War, 1914-1918* by Charles Le Goffic also published by Leonaur.

This latter by means of a note passed over my shoulder by the pilot. And here let it be said that a proper understanding between pilot and observer is one of the essential features of war flying. What the latter misses the former often picks up, for when flying at high altitudes of over 10,000 feet, field-glasses for observation purposes, with the excessive vibration of the engine, are at first very difficult to manipulate.

Our machine, one of the latest scouting types, was a beauty. She climbed rapidly and had a fast turn of speed through the air, concerning which latter feature there always seems to exist in the lay mind a deal of misapprehension, especially concerning the possibilities and peculiarities of the various types.

The aeroplane is a most curious and difficult machine to build up, because so many different factors have to be taken into consideration in the construction of it. If it be constructed for speed work, it necessitates a large engine, and hence more weight, and with its limited "lifting" capacity, some other feature has to be sacrificed, very probably petrol-tanks, thus cutting down the possible duration of flight. Similarly speed would have to be sacrificed for duration.

Thus it will be seen that an aeroplane can only specialise in one feature and cannot possess, at one and the same time, speed, lift, safety, climbing power and long durability.

The *alpha* and *omega* of the adventure was that we were within certain limits free to do what we pleased. This added a certain amount of vim and interest, especially so when compared with target-registering.

As we sail along the blue sky over green fields and steepled city, my eye constantly roams round in search of enemy aircraft, but thus far with not much luck.

The firing lines are now far behind us, and we are well over into the enemy's country. One would have thought that before now we should have encountered a stray Aviatik or so, or a patrolling Albatross.

At last! In the far distance and coming towards us at a great speed "down-wind" is a white-nosed machine, which I distinguished as "Fritz," a single tractor biplane, a hybrid of the Albatross and Aviatik types, fitted with a 225 h. p. Mercedes engine, that gives 90 miles per hour. It has a range of ten hours' flight, and carries two Maxim guns one in front, but only firing sideways, and one behind the pilot.

Immediately thoughts of an aerial combat flash across my mind. I had never taken part in one before, but had often watched them from the comfortable security of *terra firma*: during that first moment I had a bad attack of "cold feet."

A vision of many a hard-fought battle in mid-air came before my eyes. With the opposing machines darting above and below one another like two great birds, the sun glistening on the whitened planes as they turned and twisted, while all around and silhouetted against the deep blue sky were the little black and flame patches of the bursting shrapnel, it was a gloriously fascinating sight.

The uncertainty held one spellbound. Suddenly one of the machines would put down her nose and descend like a stone to earth; for a moment one's heart was in one's mouth until she would right herself and climb up again into the fray. Sometimes these wonderful battles would last as long as forty minutes or an hour, until one or the other would crash down thousands of feet to the earth below.

In a warfare of long-ranging artillery, and the scientific slaughter of an invisible foe many miles away where hand-to-hand combat was practically unknown, these duels in mid-air were a delight to friend and foe alike, for they, and they alone, were favoured with the old-time romance of war, daring and adventure.

Men in the trenches would leave their rifles, forget the enemy, and gaze with wide-open eyes at what was going on overhead; drivers of ammunition-wagons would pause on their way in the middle of the road craning their necks, the while red-hatted staff-officers would order their cars to be stopped until the fight was over.

Those two little black specks, suspended thousands of feet above were the cynosure of all eyes, and when the stricken machine came low enough for her nationality to be distinguished, if it were a black cross on either wing a shout of sheer joy would burst forth from many an anxious heart; if on the other hand, it were the three circles of red, white, and blue, a sigh would go down the lines like the rustle of the wind through the trees.

She is almost up to us by this time. I let fire with the machine-gun, but she is still beyond range. Oh, those moments of expectation! Would she fight or turn tail and run?

She elected to do the former and climbed quickly above us. Her pilot opened fire with his machine-gun. The bullets whizzed past our ears, dangerously near.

We climb in turn and lose sight of her for a moment or so. It is a complicated game of blindman's buff. We got up with her at last and both let off simultaneously. There is a language spoken in that act, a language that has neither stops, commas, letters, characters, notes, nor images. It is the language of unbounded hate. Hate to the death. We got above her and "down-wind" this time. Luck is on our side. Another tray of cartridges for the gun quickly! That's got her. She drops sharply. Her pilot must have been hit and lost control of his "joy-stick." We are right on top of her now and let the whole tray of munitions off into her back.

Suddenly down goes her nose. She rushes earthwards with a very fair speed to waft her pilot to paradise. Faster and faster she travels. Fainter, fainter does our view of her become!

Down below the hundreds are waiting anxiously, already glorying in the prize. She's down at last!

Most thankfully we turn home.

CHAPTER 17

A Great Raid

Somewhere in the North of France,
Monday.

As I walked across the aerodrome, the feeble rays of the young moon were dying in the west. It was 4.30 in the morning, with an icy-cold nor'-wester shrieking through the tree-tops, and I was very thankful that I had taken the precaution of clothing myself warmly in a wool-lined leather coat and trousers, a pair of long gum boots—invaluable for keeping out wet and cold alike—a woollen balaclava helmet under my leather aviation cap, and two pairs of gloves to keep my hands from freezing.

We had received our instructions the previous night. Ten bomb-dropping aeroplanes were to be convoyed by two battle-planes.

It may be mentioned that a bomb-dropping machine is usually of the fast, scouting variety, with a speed of well over ninety miles per hour, and is a single seater—that is to say, it carries no observer. The reason for this is not very far to seek. With two men and a machine-gun aboard, very little power remains for a supply of bombs; without an observer and a machine-gun, the bomb supply may be doubled. And the more bombs aboard the more damage can be done to the enemy.

The battle-plane is either a "pusher" (with the propeller at the rear) aeroplane, mounting a large gun at the prow, or a Caudron with two engines. Its principal duty is to protect the bomb-dropping machine from attack by enemy aircraft.

The two battle-planes were the first to get away from the ground and the others soon followed. When they had all reached an altitude of 5000 feet, they took up their pre-arranged formation with one of the battle-planes on either wing; then turned their noses eastward towards

77

the sun, and set off in the direction of the enemy lines.

Far away across the sand-dunes there came the first rays of the rising sun, casting a thousand scintillating gleams across the sea. Out in the channel was a fleet of fishing smacks, heedless of the drifting mines, bowling along merrily before the breeze to their accustomed fishing-ground. The dull grey lines and the smoke-belching funnels of a British destroyer, full out at thirty knots showed as she churned the seas into masses of white foam, leaving in her rear a long white wake. Dotted here and there' were small tramp-steamers and cargo-boats. By the sand-dunes off the coast was a long dark shape, which might easily have been mistaken for a whale, had it not been for that tell-tale periscope. It was one of our own submarines. Away in the distance was a dark irregular line, which later in the day and in a stronger light, would reveal itself as the shores of old England.

A glance at the altimeter—the instrument for registering the height—revealed the fact that we were now 6000 feet. Still climbing, the course was set further out to sea, to avoid as much as possible the anti-aircraft guns at Westende and Middlekerke.

Things ashore now began to brighten up. Along and behind the firing line there was the occasional flash of a heavy gun, followed almost immediately by dense clouds of white smoke. Along the roads there crawled, ant-like, the long columns of supply and ammunition wagons. Sometimes a big gun appeared, hauled along by a puffing traction-engine; sometimes a battalion or company of infantry or a squadron of cavalry moving up to the front line.

Running south and east were the two dull grey straggling lines of opposing trenches, so close together in places that they appeared to run into one another. We were gradually drawing nearer to those much dreaded lines where our real troubles were to begin. Already far up along the coast, it was possible to distinguish Middlekerke and the Ostend railway station.

The first anti-aircraft shot! A long-drawn-out hiss and a violent explosion in unpleasant proximity—a pretty enough exhibition to watch from the safety of *terra firma*, but deucedly uncomfortable when one is playing the leading part in the little drama. It is the first shot that is always the most unpleasant and the most terrifying.

For the next few moments there continues a fairly strenuous bombardment, which necessitates rapid climbing and diving to continually alter the range. Then the firing ceases for a short while, and all is normal again.

From behind a small wood there comes floating gaily up aloft the long and ugly shape of a "sausage" (captive balloon). Now is our chance for a little just retribution. But, apparently the Germans have seen us, for the "sausage" is being brought rapidly down towards the earth again. The temptation is too strong for two of our men, who, despite previous orders to the contrary, try their 'prentice hand with a few bombs, without success. It is easy to see that this is their first time across, for the "sausage" is the most difficult of all targets, and very rarely hit.

My map now reveals to me that we are over Ostend. More shrapnel flies up, interspersed here and there with high-explosive shell. One can feel a certain contempt for shrapnel in mid-air. The conditions are entirely different when firing across the land, than when firing straight up into the air. In the latter case the resistance is more than treble, with the result that, by the time the shrapnel reaches anything of an altitude, the best of its driving force has been expended, and bullets rattle harmlessly against the wings of the aeroplane. In fact, on one occasion a Royal Flying Corps pilot returned from a reconnaissance trip with 365 bullet-holes in various parts of his machine, which was still air-worthy.

High explosive is another matter. If it bursts reasonably near the machine, there is not the slightest chance of ever reaching the ground again in a whole condition, and even when bursting at a distance it is apt to give the aeroplane a nasty jar and sometimes upsets it entirely.

One machine has had to drop out and has turned back towards the lines, and now there are only eleven of us. More shrapnel and yet more; much too near on the last occasion. We climb rapidly higher to 10,000 feet. It is a fine, clear day, and everything beneath us is quite distinct. Even so, it is a very difficult matter to manoeuvre the machine and to use one's glasses at the same time.

One peculiarity in atmospherical conditions on the Continent is that the weather is either too misty for flying, or so remarkably clear that the airman can reconnoitre from much greater heights than in England. For the first two hours after sunrise there is invariably a heavy ground mist. Yet early morning and late afternoon are the more favourable times for flying purposes.

Ghistelles looms into view, far away to the south and bathed in a sea of light mist. It is the great German aeronautical centre in Belgium. All the large enemy raids are organised and planned at this centre. The town itself is of no great size, but it has good lines of communication by road and rail, both to the firing line and the distant bases in the immediate neighbourhood of Brussels. There are some forty hangars there, and until quite recently there were two large sheds. Probably no other spot within the German lines is so well and plentifully supplied with anti-aircraft guns as is this place.

Far away in the distance, and coming "down wind" at a very great pace, is a minute black shape, at present no larger in size than a man's hand.

An enemy machine! excitement rules high. He cannot have seen us, for no Hun airman would dream of taking on so many of our machines single handed.

Nearer and yet nearer he draws. Suddenly he sees us. He turns quickly, but is too late. Our battle-plane on the extreme right is after him. The enemy skirts the fringe of the dark cloud that hang across the horizon. After him goes our battle-plane. For a short space of time both are hidden in its depths. Then, from the distant end, there descends rapidly a small black object.

Is it British or enemy? Down she goes; a steep volplane turning into a spiral, and finally into a murderous-looking nose-dive. Thank Heaven, it is the enemy machine. I have seen the black cross on the tail. Back comes our machine triumphant, and we continue on our way to Ostend.

There are various objectives of an offensive through the air. There is the attack on enemy aircraft. This is hardly a matter for an organised raid; it is rather the errand of a cruising battle-plane. Next there comes the destruction of material; ammunition columns (usually situated in woods), parks of transport, railways, and all appertaining to them, and especially bridges and trains, stations and sidings, enemy headquarters, aeroplane and airship sheds, petrol depots, and gasworks.

Lastly, there is the bombing of troops. This is a comparatively simple matter, the best occasion on which to attack them being when they are crowded in roadways and similar areas.

Zeebrugge was at last almost within reach. The place is recognisable from the long jetty running in a large curve far out into the sea. Proceeding in a westerly direction are numerous heavy troop-trains, and standing in the sidings several locomotives with steam up, all of which incidents point to the movement of a large number of troops. In the harbour are four destroyers and three submarines. The more the merrier!

Gradually we draw nearer. It is now possible to see something of the panic in the streets and roadways. Motorcars are darting out of the city in all directions; the destroyers are hurriedly trying to make for the open sea. The anti-aircraft guns begin to open fire from every quarter. And then we commence to drop our bombs. Down they go, those ministers of death and destruction, to their targets. Huge columns of living flame leap up skywards hundreds of feet into the air. The din of the engine resounds upon the eardrums until we begin to wonder if we shall ever be able to hear distinctly again. But down below, where the guns still pound away unceasingly, the crash of the bursting shells, the violent explosions of the dropping bombs; all are strangely noiseless. It is a veritable inferno of death and destruction.

The roof-tops of the city are covered with great rolling clouds of thick black smoke. It is now almost impossible to distinguish any landmark on the ground below.

Two of our machines have already gone crashing down. The sight of them falling is the greatest shock to the nerves imaginable; it is the true test of bravery, for one always feels tempted to give up and follow them, but only for the passing second. The lust of battle grows strong again; more bombs and yet more are dropped on to the stricken city. The flying of the machines is marvellous to behold.

Another of our craft is hit, making number three; she, too, disappears into the mist beneath. Our bombs are now all exhausted and we turn thankfully homewards. Another machine drops out, to land safely on the foreshore, and, as we afterwards learn, the pilot is made a prisoner. Then we reach our own lines once more and are safe.

Chapter 18

A Daydream

Somewhere in the North of France,
Saturday.

The other day I had a dream; at six o'clock in the morning, at 10,000 feet up in the air, with the biting cold wind whistling by my ears. On all sides stretched the air, a boundless infinity; beneath, a moving panorama of wood, river, and hill, of men, guns and battlefield. Far in the distance, the waters of the North Sea glinted blue in the early morning sun; when suddenly the air became filled with a strange purring sound, and from all sides came flying hundreds of aircraft of varying shapes and sizes. Among them I noticed one, a leviathian. A long cigar-shaped, silver-tinted, super-airship; beneath and swinging easily in the breeze, the hull was in the shape of the old-fashioned sea-going steamer. For'ard was a wide expanse of promenade-deck, where could plainly be distinguished the passengers walking to and fro.

In the centre, on a raised dais, a band, resplendent in blue and gold, were strumming some popular air. Amidships a great bridge, where the officer the watch and the quartermaster were directing her course. Astern was another wide expanse of deck, but this apparently was reserved for the crew. Now a large group of men were busily engaged round a small, bullet-shaped aeroplane. With a whirr, she started off across the wide deck, and a second later was gracefully clearing the great ship's side, and missing a green and white balloon buoy literally by inches, sank rapidly in a southerly direction; and then our wireless telephone rang.

It was the big ship speaking us, "Had we seen anything of the home-bound mail?"

"No, we had not."

"Could we say what the Siberian weather conditions had been the day previous?"

"Well, nothing extraordinary, slight haze over North China."

"Strange, the Menelaus left Canton yesterday, should have reported Bombay this morning, Moscow reports her two hours overdue."

"No, we have seen nothing of the missing liner;" and, leaving the great pleasure ship miles in the rear, we skim across the Carpathians, speaking two Serbian cruisers on patrol duty along the Northern Frontier. From thence we run into a storm, have to climb to 5000, and by the time the mist and darkness clears away, the North Sea has loomed into view. Now we are more in the beaten track, swarms of small pleasure craft go cruising by; the Paris-London way is chock-a-block with traffic: cumbersome great four and eight engined merchant vessels, slim graceful pleasure craft, government vessels, two giant American liners, and an Australian non-stop mail-boat, some naval craft and small police patrol craft, endeavouring to order the converging lines, and two military transports bringing home leave men from Abyssinia.

The Far East fleet, flying majestically and impressively along with the flagship *Twentieth Century* leading the line, the hind portion tapering off gracefully and far into the rear to the smaller aeroplane—torpedo craft. The air is full of the crackling of the wireless, every master endeavouring at the same time to engage a berth in either the London or Norwich aerodromes. Soon the fleet makes a sharp turn to the left, the less important and smaller craft scurrying hurriedly away to give her passage. The Home Fleet looms into view, silent and majestic; in the dim distance the two units sight each other, and after paying the usual compliments, pass on their respective ways. Nearer the English coast the air swarms with pleasure vessels, elegant and tiny airships float lazily in the air, their occupants lolling idly in the sun. Over Dover can be seen the ugly form of the new floating dock, said to be large enough to accommodate even an air dreadnought.

Strung across the North Sea; about 2000 feet up, and well below the level of the trade routes, are the small grey ships of the Aerial Sporting League. We speak one of them. There is to be an international race this morning between London and Petrograd. Amused, we watch the long grey line at the starting-post, among the green fields of Kent, presently they are beneath us in a long extended line, two machines of our own red, white and blue, well to the fore.

We give our number and business to the Patrol airship at the Nore, and come down slowly to pick up our landing stage, somewhere east

83

of Greenwich, when suddenly the waters of the Thames below are cleft in twain, as if by an earthquake, and from the disturbance there rises a squat, peculiarly shaped craft, that commences to glide along the smooth surface of the water towards Purfleet, where she climbs gently out on to the far bank, into a wide grey slipway, some quarter of a mile in width. Still crawling along on her belly, she reaches the government repair works where, taking fresh supplies aboard, she suddenly sprouts two wings and commences climbing up into the air. Again there is an unpleasant purring noise, and a yet more unpleasant concussion. . . .

"Shrapnel," my observer bawls into my ear, "better go higher," and we do.

CHAPTER 19

A Mid-Air Battle

Somewhere in France,
Friday.

It was a sleepy old-world town hidden away in the sunny hills of Northern France, with a broad highway leading from the town in either direction and easily distinguishable from the air as being a first-class or main road, by its extraordinary width and the superabundance of traffic passing to and fro. We were still flying low and could easily distinguish the long strings of motor cars, convoys of ambulance wagons, supply and ammunition columns. In one place a battalion of reinforcements, marching up towards the firing line with their transport wagons in the rear. Further up and nearer to the firing line were a string of motor 'buses, crowded outside with Tommies, their bayonets gleaming silver as they caught the rays of the early sun. In another place a small traction-engine was hauling a chain of limbers, on which were the parts of a "grandmother" (naval 15-inch gun) being hurried up to take part in that murderous duel along the lines.

We are now getting nearer the dreaded area, and for the sake of comfort and safety have to climb higher. The surface of the earth, however, stills remains distinct. The long grey winding lines of trenches stretch away to the north and south as far as the eye can reach. In some places as much as half a mile divides them, in others they are so close together, that from above they appear to "kiss." But our happy soliloquising is broken by the burst of a shrapnel shell in the near vicinity. No more time for thought now.

A Soft Job

Diving, climbing, banking, anywhere to get from those awful shells, and who can give description to the dreadful sensations one under-

85

goes the first time under shrapnel fire in mid-air? Heaven and earth seem to be rent in twain by those murderous little balls of smoke and flame and lead.

One's past life rises before one's eyes, sometimes most unpleasantly. Shells burst all round, above, below, to the left, to the right. At one moment over the nose of the machine, the next beneath the tail. Once hit, and the aeroplane and its occupants will plunge down to an agonising death on the ground, many thousands of feet below.

"And this," once remarked a cynic of one of the flying Services, "is what the men in the trenches call a soft job."

By the time we have the opportunity of looking over the side again, we are well into the enemy's country. In appearance this is an almost absolute replica of the area behind our own lines. There are the reserve trenches; there the big- gun emplacement and the advance hospitals, battalion, brigade and divisional headquarters, and far, very far, in the background, the German G. H. Q.

An Enemy Machine

We keep a wary eye open for movements of troops or supplies, but there is nothing doing. The enemy, like ourselves, is browsing on this beautiful September morning. Again we are troubled with the bursting "Archies," and again we climb higher, this time above the clouds, that stretch all round and beneath us in a billowy snow-white sea. Slowly we creep round a big white fellow towards the sun, when out from a distant corner, like a spider from his web, there darts an enemy machine. Has he seen us? For a moment he keeps on his way, then suddenly round goes his nose, and he comes towards us "downwind" at a great pace. As he draws near we discover that he is double-engined and mounts two machine-guns. He has the advantage both in the matter of guns and speed, which counts for a great deal in an aerial combat. With a faster turn of speed and the wind at his back, a good pilot should be able to overcome an enemy machine, however large and however heavily armed.

While still about five hundred yards away, he opens fire, but without effect, his bullets fly wide on either side of us. We reserve our fire. Now he is almost on top of us, and in the upper berth, thus having a great advantage. He is over us; the great shadow of his machine comes between the sun and ourselves. All the time his observer is firing wildly, some of his shots have punctured the wings, but thank God, none came near the body. The danger is over. It has been a narrow escape.

We climb as fast as possible, then turn to find him coming to meet us, almost on end. Another machine-gun duel between the observers. We have got him this time; he is hit, he drops suddenly. A few more shots from our gun and it will be all over with him. But our gun has jammed, hastily the observer tries to remedy it. It is too late. We have missed our opportunity. Nothing else for it but to put a new tray of ammunition in the gun and have another go at him. How difficult this is in mid-air! In the safety of *terra firma* it is the easiest thing in the world to take the gun to pieces, or to change the ammunition tray, but here, in the confined space of an aeroplane up in mid-air it is an entirely different matter. We are only just ready when he turns to meet us. Another duel—he has passed by.

Again we both wheel to the combat. This time he is on top of us. We give up hope, and prepare for the worst. On the top of us again; his shooting is bad, but he has got the observer in the arm. Turn round to escape—no combat possible with the man at the gun *hors de combat*; but the observer, plucky fellow! does not know the meaning of defeat. He signals to his pilot to carry on. We turn again. The enemy is confident that he has winged us. Too confident! We wait till he is almost level with us before we fire. Then *zipp, zipp, zipp*, he is hit. He plunges downward. We get on top of him. Another round of lead into his back. It is all over, he plunges headlong to earth; and with a feeling of regret for our gallant foe, who fought so well, we turn homewards to earth, peace, and safety.

CHAPTER 20

A Battle From Above

Somewhere in the North of France,
Thursday.

Dawn—not as we imagine it; but a dawn with God's clear Heaven filled with every winged messenger of death. The very earth is shaken with agony, and the face of the sun is blotted out by heavy, choking clouds of picric smoke that hangs and hovers over the earth like a pall.

Far in the background rises a battle aeroplane. Nearer and nearer to the line it creeps, and without any attention from the enemy's anti-aircraft guns. The German artillery is too much engaged in work of a more serious nature—the work of hurling back the irresistible lines of British infantry.

The frail craft passes over the lines, and meeting with no opposition sinks lower in long, sweeping circles, and finally appears to hover, as nearly as an aeroplane can hover, some two miles to the east and well over the enemy's country. Then it is bombarded on all sides with "Archibalds," now above, now below, now immediately in front, now immediately behind, but the machine continues to manoeuvre as if entirely oblivious of shell fire. Other swiftly moving shapes have now crept out from the direction of the British base, and all are hovering over different portions of the long line of muddy trenches, while the battle rages in all its fury.

All the varied operations of the extensive battlefield are as an open book to the watch in that frail craft . . . . the battle swaying backward and forward from trench to trench, the hand-to-hand combat in the open, the ding-dong artillery duel, and the hurried rush of supports and reinforcements. Nothing can be hidden from this peering eye above, that transmits the news by wireless to the great guns far in the rear, and to the headquarters, where the commander traces every movement of the battle on his map, like a chess-player planning his moves and counter-moves on a chessboard.

The enemy's country is more heavily wooded and more broken than our own. Dotted here and there are small straggling villages. To the north, on either side of the road, are two small villages, now a mass of ruins. Between them is the tall chimney of a sugar factory, from which the black smoke no longer rises; and behind it, nearer the firing line, the long, ragged arms of a windmill move furtively in the slight breeze. To the south, and immediately in the rear of another small village, there is a large and straggling cemetery.

Woods, farms, a broken and distorted railway line, another factory, and a narrow winding stream, and the picture is complete. No! Not quite complete. Standing far removed from the main road is a large and densely wooded forest. The observer watches anxiously the stretch of British trenches immediately facing the wood. Then the barren, shell-swept land between the opposing trenches springs into life. Men and more men come swarming across the trenches and make for the German lines.

The observer watches anxiously the stretch of British trenches immediately facing the wood. There is a strange, unaccountable feeling in the air that, were it not for the never-ceasing roar of the aeroplane engine, would be hushed and silent as the moment prior to the start of a horse-race, when an element of overstrung expectancy pervades the human brain. Down below there, the Lilliputian figures crouch like ants behind the mudbank, waiting for the dread signal when the race shall commence, the race of human life and death. The booming of the great guns in the rear has long since ceased, and the nebulous region of No-man's-land, were it not for the battle-scarred earth, would resemble an ordinary peaceful country-side, so quiet and deserted has it become.

The minutes tick slowly on and on. Now it must be getting very near the appointed hour. Will it never come? Restless movements are evidenced in the opposing trenches, where an occasional bayonet glitters in the sun, or strange figures wander to and fro. At last! With a shout and roar, they are over the top. The earth trembles. Then the barren shell-swept land between the opposing trenches springs into life. Men and more men come swarming across and make for the German lines. The scene now baffles all description, it is like a fleeting glimpse of Dante's Inferno, as if all the hate and murder and courage and strength of human existence had met in one protracted struggle of life and death between savagery and civilisation. The two opposing masses intermingle, so that now it is no longer possible to distinguish each from each.

At last there comes a lull in the battle, and the aeroplane pilot, his hazardous expedition concluded and at a sign from the observer, thankfully turns for home, leaving behind him a scorched and scarred earth from which the smoke rises continuously in curling white-grey clouds.

CHAPTER 21

A True Story of the War (Being Part of the Diary of an Inhabitant)

Somewhere in Belgium,
Sunday.

Sunday again, but hardly to be imagined in these troublous times and places, with adventure for one's bedfellow, war for one's profession, and bloodshed and horrors for one's constant reflection. Despite all this there exists, and must always exist in every war that peculiar intermingling, that strange blend of horror and sentiment, hate and romance, that mixture of dross and gold. The feelings and actions that bring out all that is the most savage, the most primitive in man's nature, at the same time endowing him with the tenderness and unselfishness of a woman, the courage of a hero, and the fortitude and forbearance of a saint. Romance! I have a most charming instance to give to you my dear M——.

We met him first in December, 1914, in the little old-world town of S——. In fact I had the good fortune to be billeted upon him. The better class, or rather all those inhabitants who could afford it, had fled from the town at the first advance of the Hun hordes. But he had elected to risk his neck, and stay to comfort, and if possible to protect, the women and children. He was a queer old character was Père Dreyfus; he had lived in the little town now thirty years, since he came there first as a stripling curate. His curling brown hair had turned to an austere grey, his cheeks were hollow and shrunken, and his old back was bent almost double with shouldering other people's burdens.

By the general population he was almost idolised, men, women and even small children brought their troubles to Père Dreyfus, and they never went away without receiving the closest attention, and the warmest sympathy. As they loved and idolised him, so he reciprocated their feelings, and never tired of talking of them, in the long dark evenings, when we had the pleasure of sharing his company over a glass of old port, with *Monsieur le Maire*. He would relate vividly and with force, how in the great advance, the Uhlan patrols had ridden into the town, camped there for thirty-six hours, then returned the way they had come without, strange to say, molesting any of the population. But there was one thing that Père Dreyfus did not believe in, and that was the air.

"Bah," he was wont to say, with a contemptuous snap of his bony fingers; "mere playthings, toys, those air-machines, toys that will be shot down before they have been in the air for half-an-hour on end." He had incidentally never seen an aeroplane in flight, and little did he guess how those mere playthings were to affect his own life.

The cold, dreary winter had blossomed forth into glorious springtide, when again I came to S———. The old town had not changed much; if anything it was sleepier and drearier than ever. My first visit was to the little corner house by the great stone church; but the little corner house was no more, in its place was a pile of shattered masonry. With vague misgivings I sought M. le Maire, and found him in his stuffy, dingy little office in the Hotel de Ville. He was poring over some musty documents as I entered, but immediately left them to shake me effusively by the hand. "But where is Père Dreyfus?" I demanded of him. Where? He gave that impressive shrug of the shoulders peculiar to the Latin, and rolled his eyes meaningly towards the heavens. "Dead?" I exclaimed. "How did he die?"

"Ze airplanes," he replied; "how you call them? Ze flying machines come one night, and drop a bomb. When I go search in ze morning, ze worthy Father is no more."

Thus briefly, in as many words he recounted another tragedy of this awful war. Fortune is, indeed, a fickle jade. It had been her will that . . . . But there, the story is best told in the worthy Father's own words. I quote extracts from a little diary that it was his habit to keep, and which was all that now remained to enable us to glean a true glimpse of the old Father's personal feelings in the matter.

Monday.—The incessant thunder of heavy artillery the whole night long. Thus it has been for the past fourteen months, night after night without a break. I notice it no longer; it has become part and parcel of my everyday existence. Up at ——— yesterday those devils shot Meurice. For what reason I have not yet heard. I wonder what has become of his wife and two children? God help them if they are in their hands! Yesterday as I walked from —— I noticed high up in the sky three black specks coming over in a north-easterly direction. Our soldiers said they were German aeroplanes, but they passed away again without attempting to drop any bombs. It is not these things that we fear, but those fiendish 17-inch shells, which come over sometimes in the middle of the night and tear away a street of houses, killing, wounding, maiming. Unhappy Belgium.

Wednesday.—No change! *M. le Maire* asked me if I would billet two British soldiers today. I found them pleasant fellows enough; young lieutenants of an infantry regiment. Such youths, one of them cannot be more than eighteen years of age: a handsome boy, with the deep blue eyes and fair curling hair, typical of his race. They appear to regard the war more in the light of a big picnic. But they have not yet been up to the firing lines, nor seen the terrors of battle. Again to-day two enemy air machines came over. They hit Laroche's wine store and killed him and his wife and children. Nevertheless, I cannot help thinking that they are but of minor importance when compared with those diabolical shells.

Thursday.—The two soldiers left again this afternoon, smiling and joking as they came. All the afternoon and far into the night the infantry have been marching past, along the road, thousands of them, regiment after regiment, with their bands playing gaily at their head. The men all happy and contented, marching as if they were going on parade, instead of up to the firing line, many of them never to return. They have brave hearts these English! Many wagons of ammunition have been placed in the wood behind this house. They call it an ammunition park. Why, I know not.

Friday.—All today it rained and thundered. Thundered as if God in His Heaven were venting His wrath on the warring world below. For one long day there has been no booming of those awful guns. The road has become bare and deserted. In the evening came men into my house from the ammunition wagons in the wood. They told me that they had caught a spy. I am not surprised; this district swarms with them. But what otherwise can be expected if, previous to the war, the entire business relations of the neighbourhood were conducted with the Germans?

Every purchasable article from a motorcar to a needle was supplied from Berlin. This man was discovered in a deserted part of the wood, sending messages on a telegraph key. A sapper of the engineers saw the wire laid across the ground, and curious to know whither it led followed it along until he discovered this man. He will trouble us no more. But the unhappy result of it is, they say he signalled the position to the enemy, who will undoubtedly bombard us when the weather becomes fine again.

Saturday.—A fine clear morning. I hoped that the words of the sapper would prove themselves to be incorrect, and so they were to a certain degree. Anxiously I awaited the bombardment, and it must be confessed with a great misgiving in my heart. Ten o'clock! Eleven o'clock! Twelve o'clock! And still they did not open fire. But just before one a German Taube flew over. Unlike the air machine in the previous visits it did not fly away immediately, but came gradually lower in long sweeping circles, until with my glasses I was able to distinguish the two black crosses on the wings. Then the pom-poms began to bark and screech, and the heavens all round were marked with small white clouds of smoke no bigger than a man's hand in size, and fascinating to watch. He was a cool fellow, the pilot of that air machine: undismayed by the bursting shrapnel he continued to circle round overhead, as if taking the exact bearings of the ammunition camp.

Monday.—I was roused from my bed by a series of violent explosions. It is that infernal bombardment come at last, I thought to myself. But no! The air above was filled with a loud hum as of a hundred motors. I looked above me to find the face of the sky darkened with aircraft, all of them with the black cross on either wing; from all sides they appeared to be circling in. And every moment there would be the unpleasant rush of the falling bomb. A shattering explosion. A burst of flame! And the yell or cry of the dead and dying, the heartbreaking neigh of a wounded horse, the crash of falling timber. The series of smaller explosions as the ammunition and cartridges went off. For ten awful moments this continued, bomb followed bomb, explosion followed explosion, shrieks, cries, groans. It was a living hell. My God, these aircraft are more to be feared than those infernal guns. I—I—

Here the old Father's narrative ends, and across the page were two dull brown splashes, that tell their story but too plainly.

CHAPTER 22

Heroism In the Air

Somebody censored was engaged in a long reconnaissance trip into the enemy's country, and had already turned home when a shrapnel shell burst immediately beneath his aeroplane, smashed part of the body of the machine, and shattered the pilots leg. Rendered unconscious, he lost control, the aeroplane began to nose-dive to the earth, and fell 5000 feet. From this point the observer takes up the story:—

I had given up all hope, the earth seemed rushing up to meet us, and I prayed that our agony might not be prolonged. I shut my eyes and waited for the final crash, when, wonder of wonders, the machine began to right herself. Hardly daring to believe my eyes, I looked to the pilot's seat. The headlong rush through the cool air must have brought him round, and he was making strenuous efforts to regain control.

Luckily the enemy had given us up for lost, had ceased to shoot, and we immediately began to climb again. Then the Germans opened fire, and we only escaped with our lives through the superb pilotage of L——, with one leg shattered and blood flowing in streams. At 8000 feet he again seemed to be sinking. I hastily scrawled a note urging him to descend. He read it, shook his head decidedly, pointed to me with a smile in his white drawn face, then pointed in the direction of our lines, and carried on.

At times he would faint, and then, recovering himself, redouble his efforts. At last we were over the lines, but it seemed utterly impossible that he should be able to land the machine in his condition. But he did. Choosing a large green meadow about three miles behind the trenches, he landed as gently and as easily as if he had only been up for a practice flight, brought the machine to a stop, and fainted dead away.

This gallant pilot, as he lay mortally wounded in the field hospital, and knowing that he was dying, thought only of the terrible time his observer must have had. Thus he wrote to his mother in England:—

Mummy Dear,
Don't be alarmed at my little escapade; will be all right again soon and be with you .... Poor P——, what an awful time he must have had after I fainted and we were nose-diving headlong for the ground!
P. S.—Please don't go talking about this business to all the old dowagers of your acquaintance.

Officer R—— M—— was on a bomb-dropping and reconnaissance expedition in the neighbourhood of Y—— in the late summer of 1915. When twenty miles from our lines he was hit by shrapnel and mortally wounded in the thigh, but making up his mind not to be taken prisoner, he kept bravely on, crossed the lines, and disdaining to take advantage of the cover thus afforded and land in the first available spot, kept resolutely on to the aerodrome from which he had set out, though losing blood rapidly and knowing he had not long to live. There he made a beautiful landing, handed in his report, and fell unconscious, never to come round again.)

★★★★★★

Early in the present year an air raid was organised to bomb a town not far from Constantinople. The raid was duly carried out, but on the journey home one of our aeroplanes was hit by a shell and forced to come to earth in marsh lands beside a small river. Immediately a party of Turkish infantry rushed up to take charge of the craft, but before they could reach it another of our machines swooped down on the scene and landed close by. The pilot jumped out, ran across a field swept by Turkish rifle fire, picked up the wounded pilot, and placing him on his back, staggered across to his own machine. Still subjected to a violent fusillade, he unthrottled his engine, and with the wounded man carried before him, bravely flew off and made his own base again.

CHAPTER 23

The Evolution of the Airship

The airship is the aristocrat of the air. In jealousy and scorn the aeroplane may refer to her as "gasbag," "sausage;" may poke fun at her by reason of her unwieldy size, and laugh at her lack of speed; she still looks down on that craft with as much haughty disdain as a duchess of royal blood would bestow on a *nouveau riche*. Has she not a pedigree as long as may be forgotten?

She may trace her genealogy back to the Greek mythology and may number among her progenitors such men as Leonardo da Vinci, Cyrano de Bergerac, Francisco de Lana, Joseph Montgolfier, Blanchard, Santos Dumont and Count Zeppelin. The aeroplane is but an invention of the Twentieth Century!

Italy was the birthplace of the lighter-than-air craft; throughout the interesting history of the airship the names of famous Italian scientists predominate, and particularly those of the monastic order. Perhaps it was that convent life was inducive to study; untrammelled by the cares of the outside world, men turned their attention the sciences and developed their imaginations. Be that as it may, we find that today the Italian airships are the finest in the world, (as at time of first publication).

But although Italy may have done more than the other nations, history tells us that it was two Frenchmen, Stephen and Joseph Montgolfier, who were the first to bring the lighter-than-air craft prominently before the world.

The story goes that while rowing, Stephen's silk coat fell overboard into the water. It was placed over a hot oven to dry, and watching it, Joseph noticed that the hot air tended to make it rise. The upshot of the affair was the Montgolfier balloon.

Throughout history the lighter-than-air craft has figured prominently in warfare. In the Franco-Prussian War, during the siege of Paris alone, as many as 66 balloons left the stricken city, carrying 60 pilots, 102 passengers, 409 carrier pigeons, 9 tons of letters and telegrams, and 6 dogs.

Gaston Tissandier went over the German lines and dropped 10,000 copies of a proclamation addressed to the soldiers, asking for peace, yet declaring that France would fight to the bitter end.

In the American Civil War an aeronaut named La Fontaine went up in a balloon over an enemy camp, made his observation, rose higher into the air, and succeeded in getting into a cross-current, which carried him back to his place of departure. The first cross-channel flight was made by balloon in 1785, by Blanchard, who had with him an American doctor named Jefferies, together with a large supply of provisions, ballast and oars. This weighed the balloon down to such an extent that she almost sank into the sea a few moments after starting. Ballast was thrown overboard, and she rose, only to sink again. More ballast was dropped. Then they rose into the air and eventually landed in safety on the hills behind Calais.

Having thus shortly outlined the development of the one, we will endeavour to discover the fundamental difference between aeroplane and airship. It is simply the matter of "lift" obtained in the case of the latter from the property of being lighter than air, whereas the other craft being heavier than air must obtain its "lift" by mechanical propulsion.

The airship is merely an improvement on the old-fashioned balloon: a balloon to which mechanical propulsion has been applied. Different in shape, indeed, and fitted out with many modern improvements, its flight is still governed by the same laws of "aerostatics."

For practical purposes we will divide the airship into two portions: the envelope or balloon, and the car. Atmospheric conditions influence the envelope to no small degree. The effect of heat upon gas with which the envelope is filled is to make it expand, and consequently cause the craft to rise. Cold, on the other hand, causes the gas to contract, and the craft to descend. Air pressure is another factor which must be taken into account, and this is greatest at sea-level. The greater the altitude, the less the pressure becomes, and the less pressure on the outside surface of the envelope the easier it is for the gas to expand; but this is compensated for by the fact that the atmosphere is considerably cooler at a high altitude.

There are three types of airship: the "non-rigid," in which the two portions, the car and the envelope, are entirely separate portions, being held together by means of rigging; "semi-rigid," in which the car is partly attached to the envelope, a type greatly favoured by French and Italians; and the "rigid" airship, of which both car and envelope are in the same framework. The Zeppelin is of the latter class.

Like other great airships the Zeppelin does not rely on one single balloon for "lift." Instead, the envelope forms merely the outer covering for eighteen balloonettes, which can be regulated in the matter of expansion and contraction from the control-car of one of the three gondolas below.

We have by no means yet seen these wonderful craft at their deadliest; the German pilots are extremely brave men, yet lack that initiative and dash peculiar to the British Air Service. Were the position reversed, one dreads to think what might happen to this country.

The future is all with the airship, in the role of commerce-bearing aircraft. The aeroplane and all heavier-than-air craft are of little value save as units of war, and even then their uses are infinitesimal when compared with those of the Zeppelin. And the secret of the success of the Zeppelin is that she has the "lift," double and treble the lift of the aeroplane, and is developing beyond belief, whereas, in proportion, the aeroplane develops little year by year.

Taking everything into consideration we must have Zeppelins! It is imperative for the future safety of our nation. The longer we submit thus meekly to these aerial invasions, the longer will the war go on. The German people in the past have been intoxicated with Zeppelins. Weak, hungry and dispirited, their flagging spirits have again and again been whipped up into martial ardour by the fantastic and bragging reports issued by the General Staff in Berlin. One Zeppelin raid was of more value to the moral of the German nation than two great victories on the land. The giant craft to them is more than a mere engine of warfare and destruction, it is a fetish, almost a religion; thus after every raid the bells are rung. The streets are beflagged and decorated, and the inhabitants become mad with joy.

And we must not consider the moral effects alone. From a military point of view, at the time of writing the enemy air-raids necessitate the authorities retaining numbers of valuable aircraft and many trained and expert pilots, not to mention anti-aircraft guns and their crews, which would all be of great value on the other side. Further, Germany defeated on land, and deprived of her fleet at sea, but still in possession of her Zeppelins, is a military power, and a very strong military power of the future. We, in Great Britain, have lost forever the natural advantage we once possessed of being an island. Thanks to the vigilance and strength of our navy, we have held the narrow seas with a firm hold, that so far no other nation has been able to overcome. Now we are always open to invasion from the air; and the sea, which formerly afforded us protection, is a serious disadvantage, in that invading aircraft can creep over those broad lonely spaces, and come down upon us before we are even aware of their proximity.

How can airships raids be encountered? There are three methods. The first is, by anti-aircraft artillery; secondly, by airship; and lastly, by aeroplane. The first method—that of gunfire—is extremely unreliable. This is not the fault of the men so much, nor of the guns with which they fire, but rather of the conditions under which they work. Practice with anti-aircraft guns is rare and insufficient; and the best part of the firing takes place at night at a rapidly moving object, many thousands of feet up in the air.

Aeroplanes are greatly handicapped by want of "lift"—a quality which goes far to render aircraft either useful or useless. To obtain "lift" the latter craft relies solely on the high power of its engine, whereas, with the Zeppelin, "lift" is obtained by two means: one by the envelope, which contains gas several times lighter than air; and the other, as with the aeroplane, by engine power. Thus we have double the lifting power with a dirigible than with an aeroplane, and hence double, and in actual fact treble, the war lift; and treble the amount of bombs, ammunition, and machine-guns can be carried.

The effect the enemy hopes to gain by his constant Zeppelin raids, is partly moral, partly military. To achieve the latter it is necessary that the enemy airman destroy some position or place of military importance, as a powder-factory, an arsenal, a large camp, an important railway junction, a munitions factory, a naval dockyard, an ordnance factory, or a similar area. But in very few instances have the raiding Zeppelins touched either of these places. Thus they have achieved but little military result. The moral result attempted has been to frighten and harass the inhabitants of this country until—Germany had a mental vision—they would be grovelling on their knees in the dust, begging the government to sue for peace. We have already dealt with the moral effect these raids have on their own people. By aid of lying and bombastic reports the enemy do not fail to impress—and greatly impress—neutral countries.

Some readers will perhaps remember it was after a big Zeppelin raid on this country that Bulgaria joined the Central Powers. The Germans know only too well that we do not possess large airships of our own. Suppose we did; what would be the panic and consternation caused in Berlin by the appearance over that city of a squadron of British bomb-dropping Zeppelins, and how far would it go to shorten the war?

During the last few months we have seen the Zeppelin in a more useful and more dangerous aspect, namely in the capacity of Naval Scout. Let us consider what are the main duties of a light-cruiser fleet at sea; they are of a very similar nature to those of the cavalry, namely to form a protective screen to the main body, and to advance as nearly as possible to the enemy to discover the exact disposition of his forces. In one word their main duty is scouting. In this respect the enemy went one better than ourselves. He built Zeppelins, and succeeded in accomplishing with a single Zeppelin that which in former days had required a fleet of light cruisers. Without necessarily running any risk, the giant airship at a height of 10,000 feet has a view extending on a clear day to as much as thirty miles, and some three-hundred square miles of sea surface.

What cruiser lookout can claim a perspective equal to that? At thirty miles, or twenty-five or even twenty, the Zeppelin pilot is well out of range of the enemy shells, and with his wireless instrument, which has another range of thirty miles, can signal to the admiral of the fleet when the enemy is yet sixty miles off. This view explains the fact why the two fleets have so seldom been at grips in the two years of war. The enemy, by means of his aerial scouts, must oft and again have been warned of the proximity of the British Fleet. The official account of the Jutland battle stated that the weather was dull and misty; hence the Zeppelins would have been unable successfully to perform their usual duties.

The extreme radius of Zeppelin activity is usually considered to be 600 miles out, 600 miles home, and judged from the three principal Zeppelin centres—Heligoland, Brussels and Friedrichshaven— embraces, with the possible exception of a small and unimportant portion of the west coast of Ireland and north coast of Scotland, every city, military camp, munition factory, dockyard and industrial centre in Great Britain.

CHAPTER 24

Laws of the Air

At a recent coroner's inquest on the death of a young service pilot in England, an instructor of the flying school at which he was being trained, stated in the course of his evidence that if the pilots—there had been a horrifying collision in mid-air—had only been familiar with aerial rules and regulations, the accident would never have occurred.

In this particular instance one machine had been coming down, while another was just leaving the ground. Both of the pilots were aware of the danger they were in, but neither knew the right course to pursue. Result—collision and death. Had both of them carried out the Royal Aero Club's regulation: that an aeroplane passing another aeroplane in mid-air must leave at least ten metres space between the extreme wing-tips and always pass to the right of the approaching craft, both of them would have been alive today, (as at time of first publication).

So very few of the public outside the flying world are aware that, as navigation of the sea is ordered by the Navigation Act, so is the navigation of the air by the Aerial Navigation Acts of 1911 and 1913, and by the rules and regulations of the Royal Aero Club, which latter organisation previous to the war controlled all matters aeronautical and still controls the granting of pilot's certificates.

Even in the ballooning days a charter was drawn up at a conference at Brussels, which ordained that every private balloon—that is to say, one not in the hands of the naval or military authorities—must be registered and have a name and number, which should be printed in large letters on the body of the balloon. The place of residence of the owner must also be stated, and the number and the place of origin be printed in red. Every ascent by a private person must be under the control of a state official. Government balloons, on the other hand are not expected to carry papers, but private balloons must have a copy of the official particulars and a list of the passengers. A balloon must be identified in the same way as a ship, and must carry a flag, fastened to the net half-day down the balloon, and this must be recognisable both by its shape and colouring, and be properly mounted in position. A journal must be kept and the man in charge must produce his certificate on demand.

These latter rules also apply to airships, but not to aeroplanes. These types of aircraft are too numerous to be able to identify singly, but there are many other rules to which they must submit. For instance— flying over London and similar crowded areas is prohibited; or, in the words of the R. A. C.:

> Flying to the danger of the public is prohibited, particularly unnecessary flights over towns), or thickly populated areas, or over places where crowds are temporarily assembled, or over public enclosures at aerodromes at such a height as to involve danger to the public. Flying is also prohibited over River Regattas, Race meetings, meetings for public games and sports, except flights specifically arranged for in writing with the promoters of such Regattas, Meetings, etc.

If he disregard any of these regulations, the airman is liable to a fine not exceeding £20 and suspension of his flying certificate.

The first Aerial Navigation Act of 1911 was not in reality a Navigation Act at all, but although that was its title, it was described as "An Act to provide for the protection of the public against dangers arising from the Navigation of Aircraft." The penalties attached thereto were exceedingly heavy and provided that any airman disregarding the Act would be liable after conviction on indictment or on summary conviction to imprisonment for a term not exceeding six months, or to a fine not exceeding £200, or to both such imprisonment and fine.

The act included various prohibited flying areas, mostly in the neighbourhood of arsenals, munition factories, and naval dockyards, or similar military areas.

Certain conditions were imposed on aircraft landing in this country from abroad, as that the person in charge of the aircraft, before commencing a voyage to the United Kingdom, must apply for a clearance to a duly authorised British Consular officer. He must make a written application, which states clearly the name and registered number of the craft; the type, the name, nationality, and the place of residence of the owner or person in charge, and of every member of the crew; and the name, profession, nationality and place of residence of every passenger (if any), the nature of the cargo (if any), the approximate time of departure, place of departure, the intended landing-place in the United Kingdom, the proposed destination, and the object of the voyage.

Having settled the matters of procedure, it was further added that:

No person in any aircraft entering the United Kingdom should carry, or allow to be carried, in the aircraft, any goods, the importation of which is prohibited by the laws relating to customs; any goods chargeable upon importation into the United Kingdom with any duty or customs, except such small quantities as have been placed on board at the place of departure, as being necessary for the use during the voyage of the persons conveyed therein, any photographic apparatus, carrier or homing pigeons, explosives or firearms, or any mails.

On the return journey the aircraft is not permitted to leave unless there be at least one British representative, approved by the authorised officer, on board. No photographic or wireless apparatus, etc., shall be carried, and no mails.

Foreign, naval, or military aircraft must not pass over, nor land within any port of the United Kingdom, nor the territoral waters thereof, except on the express invitation, or with the express permission, previously obtained, of His Majesty's Government.

None of the foregoing orders applies to naval or military aircraft, belonging to, or employed in the service of His Majesty.

CHAPTER 25

Aerial Combat

With every combat in mid-air some new theory is set up, some new conclusion arrived at, and as yet nothing can be definite. We may say for practical purposes that the strategical work is confined to seaplane and airship-scouting with the fleets at sea and long-distance aeroplane raids into the enemy's country; tactical work to reconnaissance trips over the neighbourhood of the lines and the direction of artillery fire. The battle formation of the aeroplane squadron is now, and will in the future be similar to that of a fleet at sea. Even now the two methods of battle are closely akin.

There are three distinct phases of aerial combat to be considered—aeroplane *versus* aeroplane, airship against airship, and aeroplane against airship. It is a difficult matter to decide which is the more useful as a fighting unit, but thus far one is inclined to say the light, high-powered aeroplane. Zeppelins and airships are for the most part clumsy and unwieldy. Seaplanes, again, are usually heavy and slow to answer to their controls.

The important factors are the lifting power of the machine and weather conditions. The property of "lift" is determined on the one hand by mechanical devices, and on the other by the balloon portion of the craft which is lighter than the air. Lift spells speed, endurance, and climbing powers, and therefore the machine with the greater lift is the better equipped for fighting purposes.

WIND AND CLOUD

Next in order of importance is wind. The engine may be giving a speed of sixty miles per hour, and the craft be flying in the teeth of a 20 m.p.h. wind, thus its actual speed would be forty, not sixty, miles an hour. Again, two enemy machines, A and B, are approaching one another to give battle. Both have a speed of 60 m.p.h., but A is flying "down" with a fifteen-mile wind at the back of him. Their relative speeds would be: A seventy-five, B forty-five, or an advantage of thirty miles an hour for A; but on the turn—the majority of aerial combats are fought out on the principal of circling and wheeling—the advantage would be transferred to B. Good pilotage is of extreme importance; the pilot who is able to get the most out of his machine and knows it best will almost invariably gain the day.

Clouds are often made great use of by pilots. Almost every day we read of a machine dashing out from behind a bank of cloud, and taking another by surprise. On the other hand, clouds may prove disastrous to both combatants, owing to the peculiar property they possess of influencing the stability of the machine.

Lift, however, is still the great factor, since the fight always develops into a struggle for the upper berth, and is usually fought in an upward direction. It is climb, climb, climb; then, with the wind at his back, a last swoop down on the enemy—taking him in his most vulnerable position—and the fight is over. Various expedients are made use of to gain this end, such as getting between an opponent and the sun, "diving" suddenly and "looping." With either aeroplane or airship it is the uppermost position that counts.

The type of craft most useful for this work is the high-engined biplane of the "tractor"—propeller to the fore—type, the machine-gun firing through the blades of the propeller. The essentials of these machines are speed and ability to climb quickly. The slower machines, with greater powers of endurance, are more useful for bomb- raiding and reconnaissance purposes.

"Lift" the Factor

Airship combat has yet to materialise. Many opinions and theories, often widely conflicting, have been put forward concerning the possibilities and probabilities of such conflicts, but nothing definite can be advanced until a battle between airships has taken place. The opinion of the majority of the experts is that an airship would be little better than useless to meet an airship, and for our own particular requirements—that is, the repelling of Zeppelin raids—aeroplanes are of more use; which brings us to the combat between aeroplane and airship.

Considering first their main qualities: the airship has great "lifting" powers, is more heavily armed, can climb at a faster rate, and has greater powers of endurance; whereas the aeroplane has greater speed, is more easily manoeuvred, and is less unwieldy.

The tendency of the Zeppelin commanders is to increase rather than decrease altitude with every raid, which renders attack by aeroplane more difficult; but, on the other hand, aeroplanes are being built which can develop so remarkable a speed that they will soon be able to climb above Zeppelin altitude. When that occurs the Zeppelin menace will end forever.

CHAPTER 26

The Air the War and the Future

Had either Orville or Wilbur Wright, when they first glided down the low sand-dunes of the Pacific shore on a frail, uncontrollable air machine, in the earlier part of this century, or Count Zeppelin, as he worked unceasingly on his giant airship, been blessed with the imagination and the gifts of a seer—what remarkable vision would have been theirs!

To see that frail glider increase and grow into a motor-propelled, double-winged aeroplane, darting through the air with the speed of a cyclone: that unwieldy airship, capable at the most of remaining for half-an-hour in the air at a time, develop into a craft, to which the crossing and re-crossing of the wide expanse of the North Sea was an everyday occurrence: to see the aeroplane climb up to 18,000 feet in the sky, to attain a speed of over 100 miles per hour, and remain in the air for hours on end. . . .

The Zeppelin originally intended to be a peaceful carrier of the commerce of the world, converted into a ship of war, with machine-guns mounted fore and aft; and with a cargo on board deadly enough to wreck the half of a city. . . .

The far-flung battle-line of Flanders, over which there creep, like great grey wasps, French, Belgian, German and British aeroplanes alike; the elongated shapes of raiding Zeppelins, darting hither and thither over a darkened London, among piercing searchlight rays and bursting shrapnel!

Yet a few years, and the shapes and structures may undergo even more marvellous change; for every talent and accomplishment, every art and science of modern civilisation will be devoted to the development of this new science of aeronautics.

The War and Aviation

One may say, without much fear of contradiction, that the war has done more towards the development of aviation, and has rendered more things possible to be done in two years than would have been accomplished in ten years under pre-wartime conditions.

It has necessitated the production of many thousands of craft of varying degrees of size and shape, and the number of factories engaged upon the production of aeroplanes, airships, and spare parts for the respective craft has trebled. For one trained and experienced aviator, in 1914, there are today, (at time of first publication), at least ten, if anything more capable, and certainly better experienced.

As a test of the durability and the capabilities of aircraft, flying under war conditions cannot be equalled, for various reasons. Firstly, manoeuvres, which in times of peace would be considered risky to life and thus avoided, must be endured daily by pilots flying over the battle area. Flying under shell-fire frequently necessitates manoeuvres, entirely unaccounted for by the constructors of the machine, which put a very great strain on the framework, wings, struts, etc. To compensate for such strain, every wire, strut, and part of the framework is constructed of a strength at least eight times greater than that of the actual strength required. Thus the weak points of the machine are discovered, also the centres at which the greatest strain takes place.

Future Types of Craft

The shape and general build of the aeroplane has not thus far changed materially from the original models of Orville and Wilbur Wright, save that the majority of the modern machines are tractors (*i. e.* with the engine in front), whereas the older types were "pushers" (with engine at the rear). The new principle has naturally both advantage and disadvantage. With the tractor engine, the machine has a great speed, and is able to climb at a much faster rate, but the inherent stability of the craft is seriously affected—by shifting the engine 80 *per cent*, of the total weight is moved from the centre to the nose of the aeroplane. To compensate for this the wings have had to be extended, and this has added considerably to the weight in aggregate. But this evil has again been remedied, by bringing the extreme ends further to the rear, and slightly indenting each wing-tip: in a word, constructing the aeroplane more and more after the fashion of a bird in flight.

Such is the peculiar working of the human mind, however, that when some new theory or substance is evolved, similar to the one in question, it is content to concentrate on the original formula, and develops that rather than apply the same principles to an entirely new formula. Thus, after some twelve years of flying, we have only four distinct types of craft: the balloon, the airship, the aeroplane, and the seaplane—the two former being very similar both in principle and shape, as also the two latter. Exception cannot be made for the "tri-plane," for that machine, with three planes, has the same shape as the aeroplane.

The principles of aero-statics, and aero-dynamics by no means confine the constructor to these two standard forms; and in the near future the aeroplane will be built on similar lines to the ocean-going liner, and the airship very much on the same principle.

Development in size and speed depend on future experimenting, and flights have already been made both in France and Russia by gi-ant aeroplanes, in which, in one case nine, and in the other fifteen passengers, exclusive of the pilot were carried at one time; while the later Zeppelins are capable of lifting to a height of over 12,000 feet, a crew of thirty odd, with a further weight of bombs and war mate-rial aboard, and flying distances of over 800 miles. Again, there are the orthropic and the ornithropic types of craft, which their inventors claim to be capable of rising vertically from the ground to a height of 10,000 feet.

Combining these principles we ought within the space of ten years to be in possession of aircraft capable of flying at over 150 miles an hour, with a cargo of many hundred tons aboard, and with a radius of over 3000 miles, able to start and land with ease in a confined space about the size of Leicester Square. The aerial landing grounds will be the flat roofs of gigantic buildings specially constructed in the centre of London. Automatic lifts will convey the passengers from the air level to the street level, where they will be deposited in electric trains, running, in all directions. Impracticable, say the critics, but so they said when Count Zeppelin and the Wrights first started their experi-ments.

PROPERTIES OF WAR AND PEACE MACHINES

There is not, and there never was, on this earth a new idea so well deserving of examination as the science of Aeronautics. The history of that science deals with the most momentous invention in the history of civilisation. No other science allures the imagination so far forward into the dim future, when the business of the world will be carried up from the level of the sea and the land to that of mid-air, and when travel will be so rapid and safe that space will almost cease to be an obstacle to man's communications.

The proudest inventions of the late nineteenth and early twentieth centuries are but of yesterday when compared with those of the aeroplane and airship. It is, therefore, of the utmost importance that we should consider how the development of aeronautics will affect the future of the human race. Under the present wartime conditions, (as at time of first publication), there exists a grave danger that the aeroplane and the airship will be developed too much for war purposes to the detriment of future commercial uses. The qualities mainly required by the war machine, speed, ability to climb quickly, and compactness, differ entirely from those required by the peace time or commerce-bearing aircraft, which have ability to remain in the air for a great space of time, and to fly greater distances. The extra speed required by the war machine may easily be dispensed with in the commerce-bearing machines, as also may altitude, for whereas the war machine must fly at a height of over 12,000 feet, a height of between 2000 and 3000 will suffice under ordinary conditions, and it will be at this altitude that the best part of the flying will take place after the war.

FUTURE NAVIES AND ARMIES OF THE WORLD

How will aviation effect warfare in the future? Will it abolish entirely this undesirable condition of affairs, or will it serve to provide added inducements? It is, indeed, a debatable point. If we incline to the latter view, every known argument and theory points to the fact that warfare of the future will be to all intents and purposes instantaneous. There will be no preparatory delay caused y the necessity of placing large armies in the field, of gradually marching forward to establish contact with the enemy, and of carrying out skirmishes which may be prolonged to weeks and months before the actual battle takes place. The belligerent fleets will set off in the dawn or in the darkness, as the case may be, and before twelve hours have elapsed, after entering into the conflict a definite decision will have been reached. For the airman, there is no falling back to a second line of trenches, to a natural position heavily defended, or to a concrete fortress or emplacement, or to fight a rearguard action. The fight in the air must be to the death, without quarter asked or given, for no prisoners can be taken. The loss of men and material will be tremendous.

It is doubtful whether aviation will entirely do away with fighting on land and sea, but it is very obvious that either fleet or army of one belligerent nation, at the mercy of the air fleet of another nation, will be in a very helpless position. Should the warfare in the air be indecisive, were such a condition within the realm of reasonable argument, it might be possible for the fleet or army to be brought into action with advantage, but even this is doubtful. As regards our own nation, before 1926, the Royal Naval Air Service will be the largest and most important service in Great Britain. Possibly there will be a single Air Service, and before ten years will have elapsed it will be the most important of all the British services, and will be composed both of aeroplanes and airships. The only other form of aircraft, the seaplane, being too slow, too clumsy, and too costly, will long ago have been abandoned.

Peace and War Uses of Aircraft

Before we enter upon the discussion which is the subject of this paragraph we wish to guard ourselves against one misconception. It is possible that readers of this chapter may already have come to the conclusion that it is possible to develop aircraft for one purpose, and one purpose only: that is, either for war or for commerce; and impossible to develop them for both. This would be an entirely erroneous idea. It is true that we have already laid stress upon the fact that there is a very imminent danger that aircraft may be developed too greatly for war purposes to the detriment of others, but provided that the necessary precautions are taken, there is yet ample time for the commerce-carrying machine to be developed at the same time and in the same manner as the war machine.

Within a very short time we may find that the Super-Zeppelin of the air will have entirely replaced not only the Dreadnought of the sea, but also the giant passenger liners. Both the war and the peace craft will be considerably larger in size than the 1916 type; the balloon portion of the Zeppelin will have trebled itself in size; it will be, if anything, of greater length and of slimmer formation, while the covering will be composed of some light but durable metal, such as aluminium, to prevent the possibility of explosion of gas caused by the firing of the guns.

The narrow gondola beneath will be wider, and will mount several guns of 4.7-inch or larger calibre: for although the Zeppelin of the future will be a much more stable and airworthy craft, by reason of its lateral stability it will never be possible to fire a gun of any size from either bow or stern of an airship or a Zeppelin, without bringing the whole craft canting over, and possibly breaking its back. Thus, all Super-Zeppelins of the future will be heavily armed amidships, that is to say, where the proportion of strain on the craft is least felt. The passenger-carrying variety will differ very slightly from the war machine, save that the gondola will be deeper, more graceful, and more on the lines of the hull of the present-day ocean-going ship or steamer. The Parseval and similar types of large airship will replace the cruiser and the battle-cruiser; also the large cargo-bearing steamers of today, (as at time of first publication).

With regard to the aeroplane, we are already in possession of super-craft, some of double engine variety, the Sykorsky, the giant Russian machine, and the triplane, or three-planed aeroplane; but it is extremely doubtful whether it is possible for the aeroplane, being a heavier type of aircraft, to develop into a much larger size than it is today; the reason for this being the abnormal engine-power that would be required to lift such a craft from the ground, and the fact that the extra weight thus occasioned would render the whole craft unairworthy. However, the aeroplane will fulfil in the future the uses of the light-cruiser and the torpedo-boat, while a sort of seaplane submarine will fulfil the double purpose of both over and under water work.

As a commercial vessel the aeroplane will only be of use for the conveyance of passengers and light cargoes on short voyages from Great Britain to Ireland, Great Britain to France, Holland, Norway, or Russia.

THE BALANCE OF POWER

The new method of warfare will not influence to any material extent the present condition of international politics. Of all the Great Powers, however, Great Britain will be more nearly affected. For many centuries past we have relied upon our natural geographical position, as an island, to protect us from all invasion. And to retain this insular and impregnable position we have relied upon our glorious navy, which is, and always has been, mistress of the seas. But now we are no longer an island; that is to say, we are no longer protected from the attacks of an enemy merely because we are surrounded by sea, even although we maintain the supremacy of our naval power. Another element has now to be considered, namely, the air, and that, unfortunately, we do not hold with the same mastery that we did the sea.

It will be seen, therefore, that for the safety of the empire, we must immediately build up a great air fleet, and gain the supremacy of the air. Germany has already shown us the lead in this respect, and we must not be content to follow, but to improve, greatly improve, upon that lead. One thing is certain, that the mad extravagant race for armaments among the nations will continue, but with this difference—that it will be for great fleets of the air, as today it is for large armies and great sea fleets.

FUTURE INFLUENCES

Thus far we have dealt solely with the influences of aviation upon warfare and upon commerce; but such influences will by no means be confined to these two phases; there are many other features in international life that the development of aeronautics will influence greatly. Foremost amongst them is that of travel. For the first few years the cost of travel in the air will be appreciably greater than is now the case. One of the leading aeronautical experts of the day has computed that, to run a commercial service of aircraft, to cover the heavy expenditure that will be incurred, and to allow for the wear and tear of machines, it will be necessary to make a charge of $1\frac{1}{2}d$. per mile, or a 50 *per cent*, increase on the rates for present day travel by steamer and railway. Once the project is in full swing, however, and the initial outlay has been recovered, such charge will be reduced to one halfpenny per mile, or 50 *per cent*, less than present conditions.

In the matter of speed and time, there will be a remarkable advantage; for example, some of the proposed air routes are London to New York in 18 hours, London to Capetown in 54, and London to Sydney (Australia) in four days. This added economy and speed will tempt the travelling public, and for that matter the non-travelling public further afield, and will serve greatly to help on education and the rapid development of the remotest of our colonies, thus drawing closer the bond of union between the different portions of our great empire. Countries and tracts of land hitherto undeveloped and unknown will be opened up by the aerial explorer, and whole continents will, with the greatest ease, be policed by aeroplane and by airship.

A FUTURE WAR WITH GERMANY

Will this war be followed by an aerial war between Germany and Great Britain at a no distant date? This depends solely on the future course and the conclusion of the present war.

After some fourteen years' experimenting, inventing and developing, and the expenditure of several millions of money, Count Zeppelin, or rather the very considerable staff of experts which he has at his disposal, produced the modern Zeppelin: that is to say, the craft that has been in use since the outbreak of the war. What Germany's policy was in constructing these huge craft it is not difficult to discover. Previous to August, 1914, when her navy was inferior to only one other in the world, and that our own, and she was gradually gaining upon us both in the number of ships and *personnel*, very little was heard of the airship programme: the industry was given State encouragement; but then, to our cost, we know that the enemy has always encouraged any new enterprise that was likely to prove of value from a military point of view. War was declared.

Our gallant fleet, by a series of brilliant engagements, succeeded in driving the enemy shipping from the seas of the world, and in bottling up the *Kaiser's* grand fleet in the Kiel Canal, where it has ever since remained. What effect did this have on the aircraft, and more particularly the Zeppelin, industry in Germany? Labour was instantly withdrawn from the ship-building yards, and turned over to the construction of Zeppelins. In the early stages of the war the output stood at approximately one a month; this soon crept up to a couple a month, then to three, then to one a week, and now today they claim that two Zeppelins per week are being turned out by the factories that have sprung up in nearly every large town in the German Empire. What do all these events portend? Those who know the German and his characteristics intimately, tell us that at the back of every German mind, the keenest of all desires is an invasion of England.

The reason for this bitter hatred is that the British Empire is on every hand an obstacle to the development of Germany; we were their keenest trade rivals, their most dangerous enemy in the matter of world supremacy, and we were successful in establishing colonies, an ambition dear to every German heart. There can only be two objects in view in the mind of the German Imperial Staff: the one is a gigantic air raid on this country, as a last resource during the present war; the other, a determination on the part of Germany, after the present war is ended and forgotten, to gain a considerable ascendancy in the air, and thus once more to take her place as a martial power among the nations. To prevent this, it will be necessary for us not only to destroy her armies on the land, and her fleets at sea, but also her fleets of aircraft; for Germany, though beaten by land and sea, and still in possession of her aircraft, will remain forever a menace to the civilised world.

LEONAUR

ALSO FROM LEONAUR
AVAILABLE IN SOFTCOVER OR HARDCOVER WITH DUST JACKET

WINGED WARFARE by *William A. Bishop*—The Experiences of a Canadian 'Ace' of the R.F.C. During the First World War.

THE STORY OF THE LAFAYETTE ESCADRILLE by *George Thenault*—A famous fighter squadron in the First World War by its commander..

R.F.C.H.Q. by *Maurice Baring*—The command & organisation of the British Air Force during the First World War in Europe.

SIXTY SQUADRON R.A.F. by *A. J. L. Scott*—On the Western Front During the First World War.

THE STRUGGLE IN THE AIR by *Charles C. Turner*—The Air War Over Europe During the First World War.

WITH THE FLYING SQUADRON by *H. Rosher*—Letters of a Pilot of the Royal Naval Air Service During the First World War.

OVER THE WEST FRONT by *"Spin" & "Contact"* —Two Accounts of British Pilots During the First World War in Europe, Short Flights With the Cloud Cavalry by "Spin" and Cavalry of the Clouds by "Contact".

SKYFIGHTERS OF FRANCE by *Henry Farré*—An account of the French War in the Air during the First World War.

THE HIGH ACES by *Laurence la Tourette Driggs*—French, American, British, Italian & Belgian pilots of the First World War 1914-18.

PLANE TALES OF THE SKIES by *Wilfred Theodore Blake*—The experiences of pilots over the Western Front during the Great War.

IN THE CLOUDS ABOVE BAGHDAD by *J. E. Tennant*—Recollections of the R. F. C. in Mesopotamia during the First World War against the Turks.

THE SPIDER WEB by *P. I. X. (Theodore Douglas Hallam)*—Royal Navy Air Service Flying Boat Operations During the First World War by a Flight Commander

EAGLES OVER THE TRENCHES by *James R. McConnell & William B. Perry*—Two First Hand Accounts of the American Escadrille at War in the Air During World War 1-Flying For France: With the American Escadrille at Verdun and Our Pilots in the Air

KNIGHTS OF THE AIR by *Bennett A. Molter*—An American Pilot's View of the Aerial War of the French Squadrons During the First World War.

AVAILABLE ONLINE AT **www.leonaur.com**
AND FROM ALL GOOD BOOK STORES

07/09

ALSO FROM LEONAUR

AVAILABLE IN SOFTCOVER OR HARDCOVER WITH DUST JACKET

ZULU:1879 by D.C.F. Moodie & the Leonaur Editors—The Anglo-Zulu War of 1879 from contemporary sources: First Hand Accounts, Interviews, Dispatches, Official Documents & Newspaper Reports.

THE RED DRAGOON by W.J. Adams—With the 7th Dragoon Guards in the Cape of Good Hope against the Boers & the Kaffir tribes during the 'war of the axe' 1843-48'.

THE RECOLLECTIONS OF SKINNER OF SKINNER'S HORSE by James Skinner—James Skinner and his 'Yellow Boys' Irregular cavalry in the wars of India between the British, Mahratta, Rajput, Mogul, Sikh & Pindarree Forces.

A CAVALRY OFFICER DURING THE SEPOY REVOLT by A. R. D. Mackenzie—Experiences with the 3rd Bengal Light Cavalry, the Guides and Sikh Irregular Cavalry from the outbreak to Delhi and Lucknow.

A NORFOLK SOLDIER IN THE FIRST SIKH WAR by J W Baldwin—Experiences of a private of H.M. 9th Regiment of Foot in the battles for the Punjab, India 1845-6.

TOMMY ATKINS' WAR STORIES: 14 FIRST HAND ACCOUNTS—Fourteen first hand accounts from the ranks of the British Army during Queen Victoria's Empire.

THE WATERLOO LETTERS by H. T. Siborne—Accounts of the Battle by British Officers for its Foremost Historian.

NEY: GENERAL OF CAVALRY VOLUME 1—1769-1799 by Antoine Bulos—The Early Career of a Marshal of the First Empire.

NEY: MARSHAL OF FRANCE VOLUME 2—1799-1805 by Antoine Bulos—The Early Career of a Marshal of the First Empire.

AIDE-DE-CAMP TO NAPOLEON by Philippe-Paul de Ségur—For anyone interested in the Napoleonic Wars this book, written by one who was intimate with the strategies and machinations of the Emperor, will be essential reading.

TWILIGHT OF EMPIRE by Sir Thomas Ussher & Sir George Cockburn—Two accounts of Napoleon's Journeys in Exile to Elba and St. Helena: Narrative of Events by Sir Thomas Ussher & Napoleon's Last Voyage: Extract of a diary by Sir George Cockburn.

PRIVATE WHEELER by William Wheeler—The letters of a soldier of the 51st Light Infantry during the Peninsular War & at Waterloo.

AVAILABLE ONLINE AT **www.leonaur.com**
AND FROM ALL GOOD BOOK STORES

07/09

LEONAUR

ALSO FROM LEONAUR
AVAILABLE IN SOFTCOVER OR HARDCOVER WITH DUST JACKET

OFFICERS & GENTLEMEN *by Peter Hawker & William Graham*—Two Accounts of British Officers During the Peninsula War: Officer of Light Dragoons by Peter Hawker & Campaign in Portugal and Spain by William Graham .

THE WALCHEREN EXPEDITION *by Anonymous*—The Experiences of a British Officer of the 81st Regt. During the Campaign in the Low Countries of 1809.

LADIES OF WATERLOO *by Charlotte A. Eaton, Magdalene de Lancey & Juana Smith*—The Experiences of Three Women During the Campaign of 1815: Waterloo Days by Charlotte A. Eaton, A Week at Waterloo by Magdalene de Lancey & Juana's Story by Juana Smith.

JOURNAL OF AN OFFICER IN THE KING'S GERMAN LEGION *by John Frederick Hering*—Recollections of Campaigning During the Napoleonic Wars.

JOURNAL OF AN ARMY SURGEON IN THE PENINSULAR WAR *by Charles Boutflower*—The Recollections of a British Army Medical Man on Campaign During the Napoleonic Wars.

ON CAMPAIGN WITH MOORE AND WELLINGTON *by Anthony Hamilton*—The Experiences of a Soldier of the 43rd Regiment During the Peninsular War.

THE ROAD TO AUSTERLITZ *by R. G. Burton*—Napoleon's Campaign of 1805.

SOLDIERS OF NAPOLEON *by A. J. Doisy De Villargennes & Arthur Chuquet*—The Experiences of the Men of the French First Empire: Under the Eagles by A. J. Doisy De Villargennes & Voices of 1812 by Arthur Chuquet .

INVASION OF FRANCE, 1814 *by F. W. O. Maycock*—The Final Battles of the Napoleonic First Empire.

LEIPZIG—A CONFLICT OF TITANS *by Frederic Shoberl*—A Personal Experience of the 'Battle of the Nations' During the Napoleonic Wars, October 14th-19th, 1813.

SLASHERS *by Charles Cadell*—The Campaigns of the 28th Regiment of Foot During the Napoleonic Wars by a Serving Officer.

BATTLE IMPERIAL *by Charles William Vane*—The Campaigns in Germany & France for the Defeat of Napoleon 1813-1814.

SWIFT & BOLD *by Gibbes Rigaud*—The 60th Rifles During the Peninsula War.

AVAILABLE ONLINE AT **www.leonaur.com**
AND FROM ALL GOOD BOOK STORES

07/09

LEONAUR

ALSO FROM LEONAUR

AVAILABLE IN SOFTCOVER OR HARDCOVER WITH DUST JACKET

ADVENTURES OF A YOUNG RIFLEMAN *by Johann Christian Maempel*—The Experiences of a Saxon in the French & British Armies During the Napoleonic Wars.

THE HUSSAR *by Norbert Landsheit & G. R. Gleig*—A German Cavalryman in British Service Throughout the Napoleonic Wars.

RECOLLECTIONS OF THE PENINSULA *by Moyle Sherer*—An Officer of the 34th Regiment of Foot—'The Cumberland Gentlemen'—on Campaign Against Napoleon's French Army in Spain.

MARINE OF REVOLUTION & CONSULATE *by Moreau de Jonnès*—The Recollections of a French Soldier of the Revolutionary Wars 1791-1804.

GENTLEMEN IN RED *by John Dobbs & Robert Knowles*—Two Accounts of British Infantry Officers During the Peninsular War Recollections of an Old 52nd Man by John Dobbs An Officer of Fusiliers by Robert Knowles.

CORPORAL BROWN'S CAMPAIGNS IN THE LOW COUNTRIES *by Robert Brown*—Recollections of a Coldstream Guard in the Early Campaigns Against Revolutionary France 1793-1795.

THE 7TH (QUEENS OWN) HUSSARS: Volume 2—1793-1815 *by C. R. B. Barrett*—During the Campaigns in the Low Countries & the Peninsula and Waterloo Campaigns of the Napoleonic Wars. Volume 2: 1793-1815.

THE MARENGO CAMPAIGN 1800 *by Herbert H. Sargent*—The Victory that Completed the Austrian Defeat in Italy.

DONALDSON OF THE 94TH—SCOTS BRIGADE *by Joseph Donaldson*—The Recollections of a Soldier During the Peninsula & South of France Campaigns of the Napoleonic Wars.

A CONSCRIPT FOR EMPIRE *by Philippe as told to Johann Christian Maempel*—The Experiences of a Young German Conscript During the Napoleonic Wars.

JOURNAL OF THE CAMPAIGN OF 1815 *by Alexander Cavalié Mercer*—The Experiences of an Officer of the Royal Horse Artillery During the Waterloo Campaign.

NAPOLEON'S CAMPAIGNS IN POLAND 1806-7 *by Robert Wilson*—The campaign in Poland from the Russian side of the conflict.

AVAILABLE ONLINE AT www.leonaur.com
AND FROM ALL GOOD BOOK STORES
07/09

LEONAUR

ALSO FROM LEONAUR
AVAILABLE IN SOFTCOVER OR HARDCOVER WITH DUST JACKET

OMPTEDA OF THE KING'S GERMAN LEGION *by Christian von Ompteda*—A Hanoverian Officer on Campaign Against Napoleon.

LIEUTENANT SIMMONS OF THE 95TH (RIFLES) *by George Simmons*—Recollections of the Peninsula, South of France & Waterloo Campaigns of the Napoleonic Wars.

A HORSEMAN FOR THE EMPEROR *by Jean Baptiste Gazzola*—A Cavalryman of Napoleon's Army on Campaign Throughout the Napoleonic Wars.

SERGEANT LAWRENCE *by William Lawrence*—With the 40th Regt. of Foot in South America, the Peninsular War & at Waterloo.

CAMPAIGNS WITH THE FIELD TRAIN *by Richard D. Henegan*—Experiences of a British Officer During the Peninsula and Waterloo Campaigns of the Napoleonic Wars.

CAVALRY SURGEON *by S. D. Broughton*—On Campaign Against Napoleon in the Peninsula & South of France During the Napoleonic Wars 1812-1814.

MEN OF THE RIFLES *by Thomas Knight, Henry Curling & Jonathan Leach*—The Reminiscences of Thomas Knight of the 95th (Rifles) by Thomas Knight, Henry Curling's Anecdotes by Henry Curling & The Field Services of the Rifle Brigade from its Formation to Waterloo by Jonathan Leach.

THE ULM CAMPAIGN 1805 *by F. N. Maude*—Napoleon and the Defeat of the Austrian Army During the 'War of the Third Coalition'.

SOLDIERING WITH THE 'DIVISION' *by Thomas Garrety*—The Military Experiences of an Infantryman of the 43rd Regiment During the Napoleonic Wars.

SERGEANT MORRIS OF THE 73RD FOOT *by Thomas Morris*—The Experiences of a British Infantryman During the Napoleonic Wars-Including Campaigns in Germany and at Waterloo.

A VOICE FROM WATERLOO *by Edward Cotton*—The Personal Experiences of a British Cavalryman Who Became a Battlefield Guide and Authority on the Campaign of 1815.

NAPOLEON AND HIS MARSHALS *by J. T. Headley*—The Men of the First Empire.

AVAILABLE ONLINE AT **www.leonaur.com**
AND FROM ALL GOOD BOOK STORES

07/09

LEONAUR

ALSO FROM LEONAUR
AVAILABLE IN SOFTCOVER OR HARDCOVER WITH DUST JACKET

COLBORNE: A SINGULAR TALENT FOR WAR *by John Colborne*—The Napoleonic Wars Career of One of Wellington's Most Highly Valued Officers in Egypt, Holland, Italy, the Peninsula and at Waterloo.

NAPOLEON'S RUSSIAN CAMPAIGN *by Philippe Henri de Segur*—The Invasion, Battles and Retreat by an Aide-de-Camp on the Emperor's Staff.

WITH THE LIGHT DIVISION *by John H. Cooke*—The Experiences of an Officer of the 43rd Light Infantry in the Peninsula and South of France During the Napoleonic Wars.

WELLINGTON AND THE PYRENEES CAMPAIGN VOLUME I: FROM VITORIA TO THE BIDASSOA *by F. C. Beatson*—The final phase of the campaign in the Iberian Peninsula.

WELLINGTON AND THE INVASION OF FRANCE VOLUME II: THE BIDASSOA TO THE BATTLE OF THE NIVELLE *by F. C. Beatson*—The final phase of the campaign in the Iberian Peninsula.

WELLINGTON AND THE FALL OF FRANCE VOLUME III: THE GAVES AND THE BATTLE OF ORTHEZ *by F. C. Beatson*—The final phase of the campaign in the Iberian Peninsula.

NAPOLEON'S IMPERIAL GUARD: FROM MARENGO TO WATERLOO *by J. T. Headley*—The story of Napoleon's Imperial Guard and the men who commanded them.

BATTLES & SIEGES OF THE PENINSULAR WAR *by W. H. Fitchett*—Corunna, Busaco, Albuera, Ciudad Rodrigo, Badajos, Salamanca, San Sebastian & Others.

SERGEANT GUILLEMARD: THE MAN WHO SHOT NELSON? *by Robert Guillemard*—A Soldier of the Infantry of the French Army of Napoleon on Campaign Throughout Europe.

WITH THE GUARDS ACROSS THE PYRENEES *by Robert Batty*—The Experiences of a British Officer of Wellington's Army During the Battles for the Fall of Napoleonic France, 1813 .

A STAFF OFFICER IN THE PENINSULA *by E. W. Buckham*—An Officer of the British Staff Corps Cavalry During the Peninsula Campaign of the Napoleonic Wars.

THE LEIPZIG CAMPAIGN: 1813—NAPOLEON AND THE "BATTLE OF THE NATIONS" *by F. N. Maude*—Colonel Maude's analysis of Napoleon's campaign of 1813 around Leipzig.

AVAILABLE ONLINE AT **www.leonaur.com**
AND FROM ALL GOOD BOOK STORES
07/09

LEONAUR

ALSO FROM LEONAUR
AVAILABLE IN SOFTCOVER OR HARDCOVER WITH DUST JACKET

BUGEAUD: A PACK WITH A BATON by *Thomas Robert Bugeaud*—The Early Campaigns of a Soldier of Napoleon's Army Who Would Become a Marshal of France.

WATERLOO RECOLLECTIONS by *Frederick Llewellyn*—Rare First Hand Accounts, Letters, Reports and Retellings from the Campaign of 1815.

SERGEANT NICOL by *Daniel Nicol*—The Experiences of a Gordon Highlander During the Napoleonic Wars in Egypt, the Peninsula and France.

THE JENA CAMPAIGN: 1806 by *F. N. Maude*—The Twin Battles of Jena & Auerstadt Between Napoleon's French and the Prussian Army.

PRIVATE O'NEIL by *Charles O'Neil*—The recollections of an Irish Rogue of H. M. 28th Regt.—The Slashers—during the Peninsula & Waterloo campaigns of the Napoleonic war.

ROYAL HIGHLANDER by *James Anton*—A soldier of H.M 42nd (Royal) Highlanders during the Peninsular, South of France & Waterloo Campaigns of the Napoleonic Wars.

CAPTAIN BLAZE by *Elzéar Blaze*—Life in Napoleons Army.

LEJEUNE VOLUME 1 by *Louis-François Lejeune*—The Napoleonic Wars through the Experiences of an Officer on Berthier's Staff.

LEJEUNE VOLUME 2 by *Louis-François Lejeune*—The Napoleonic Wars through the Experiences of an Officer on Berthier's Staff.

CAPTAIN COIGNET by *Jean-Roch Coignet*—A Soldier of Napoleon's Imperial Guard from the Italian Campaign to Russia and Waterloo.

FUSILIER COOPER by *John S. Cooper*—Experiences in the 7th (Royal) Fusiliers During the Peninsular Campaign of the Napoleonic Wars and the American Campaign to New Orleans.

FIGHTING NAPOLEON'S EMPIRE by *Joseph Anderson*—The Campaigns of a British Infantryman in Italy, Egypt, the Peninsular & the West Indies During the Napoleonic Wars.

CHASSEUR BARRES by *Jean-Baptiste Barres*—The experiences of a French Infantryman of the Imperial Guard at Austerlitz, Jena, Eylau, Friedland, in the Peninsular, Lutzen, Bautzen, Zinnwald and Hanau during the Napoleonic Wars.

AVAILABLE ONLINE AT www.leonaur.com
AND FROM ALL GOOD BOOK STORES

07/09

ALSO FROM LEONAUR

AVAILABLE IN SOFTCOVER OR HARDCOVER WITH DUST JACKET

CAPTAIN COIGNET *by Jean-Roch Coignet*—A Soldier of Napoleon's Imperial Guard from the Italian Campaign to Russia and Waterloo.

HUSSAR ROCCA *by Albert Jean Michel de Rocca*—A French cavalry officer's experiences of the Napoleonic Wars and his views on the Peninsular Campaigns against the Spanish, British And Guerilla Armies.

MARINES TO 95TH (RIFLES) *by Thomas Fernyhough*—The military experiences of Robert Fernyhough during the Napoleonic Wars.

LIGHT BOB *by Robert Blakeney*—The experiences of a young officer in H.M 28th & 36th regiments of the British Infantry during the Peninsular Campaign of the Napoleonic Wars 1804 - 1814.

WITH WELLINGTON'S LIGHT CAVALRY *by William Tomkinson*—The Experiences of an officer of the 16th Light Dragoons in the Peninsular and Waterloo campaigns of the Napoleonic Wars.

SERGEANT BOURGOGNE *by Adrien Bourgogne*—With Napoleon's Imperial Guard in the Russian Campaign and on the Retreat from Moscow 1812 - 13.

SURTEES OF THE 95TH (RIFLES) *by William Surtees*—A Soldier of the 95th (Rifles) in the Peninsular campaign of the Napoleonic Wars.

SWORDS OF HONOUR *by Henry Newbolt & Stanley L. Wood*—The Careers of Six Outstanding Officers from the Napoleonic Wars, the Wars for India and the American Civil War.

ENSIGN BELL IN THE PENINSULAR WAR *by George Bell*—The Experiences of a young British Soldier of the 34th Regiment 'The Cumberland Gentlemen' in the Napoleonic wars.

HUSSAR IN WINTER *by Alexander Gordon*—A British Cavalry Officer during the retreat to Corunna in the Peninsular campaign of the Napoleonic Wars.

THE COMPLEAT RIFLEMAN HARRIS *by Benjamin Harris as told to and transcribed by Captain Henry Curling, 52nd Regt. of Foot*—The adventures of a soldier of the 95th (Rifles) during the Peninsular Campaign of the Napoleonic Wars.

THE ADVENTURES OF A LIGHT DRAGOON *by George Farmer & G.R. Gleig*—A cavalryman during the Peninsular & Waterloo Campaigns, in captivity & at the siege of Bhurtpore, India.

AVAILABLE ONLINE AT **www.leonaur.com**
AND FROM ALL GOOD BOOK STORES

07/09

LEONAUR

ALSO FROM LEONAUR

AVAILABLE IN SOFTCOVER OR HARDCOVER WITH DUST JACKET

THE LIFE OF THE REAL BRIGADIER GERARD VOLUME 1—THE YOUNG HUSSAR 1782-1807 *by Jean-Baptiste De Marbot*—A French Cavalryman Of the Napoleonic Wars at Marengo, Austerlitz, Jena, Eylau & Friedland.

THE LIFE OF THE REAL BRIGADIER GERARD VOLUME 2—IMPERIAL AIDE-DE-CAMP 1807-1811 *by Jean-Baptiste De Marbot*—A French Cavalryman of the Napoleonic Wars at Saragossa, Landshut, Eckmuhl, Ratisbon, Aspern-Essling, Wagram, Busaco & Torres Vedras.

THE LIFE OF THE REAL BRIGADIER GERARD VOLUME 3—COLONEL OF CHASSEURS 1811-1815 *by Jean-Baptiste De Marbot*—A French Cavalryman in the retreat from Moscow, Lutzen, Bautzen, Katzbach, Leipzig, Hanau & Waterloo.

THE INDIAN WAR OF 1864 *by Eugene Ware*—The Experiences of a Young Officer of the 7th Iowa Cavalry on the Western Frontier During the Civil War.

THE MARCH OF DESTINY *by Charles E. Young & V. Devinny*—Dangers of the Trail in 1865 by Charles E. Young & The Story of a Pioneer by V. Devinny, two Accounts of Early Emigrants to Colorado.

CROSSING THE PLAINS *by William Audley Maxwell*—A First Hand Narrative of the Early Pioneer Trail to California in 1857.

CHIEF OF SCOUTS *by William F. Drannan*—A Pilot to Emigrant and Government Trains, Across the Plains of the Western Frontier.

THIRTY-ONE YEARS ON THE PLAINS AND IN THE MOUNTAINS *by William F. Drannan*—William Drannan was born to be a pioneer, hunter, trapper and wagon train guide during the momentous days of the Great American West.

THE INDIAN WARS VOLUNTEER *by William Thompson*—Recollections of the Conflict Against the Snakes, Shoshone, Bannocks, Modocs and Other Native Tribes of the American North West.

THE 4TH TENNESSEE CAVALRY *by George B. Guild*—The Services of Smith's Regiment of Confederate Cavalry by One of its Officers.

COLONEL WORTHINGTON'S SHILOH *by T. Worthington*—The Tennessee Campaign, 1862, by an Officer of the Ohio Volunteers.

FOUR YEARS IN THE SADDLE *by W. L. Curry*—The History of the First Regiment Ohio Volunteer Cavalry in the American Civil War.

AVAILABLE ONLINE AT www.leonaur.com
AND FROM ALL GOOD BOOK STORES

07/09

ALSO FROM LEONAUR
AVAILABLE IN SOFTCOVER OR HARDCOVER WITH DUST JACKET

LIFE IN THE ARMY OF NORTHERN VIRGINIA *by Carlton McCarthy*—The Observations of a Confederate Artilleryman of Cutshaw's Battalion During the American Civil War 1861-1865.

HISTORY OF THE CAVALRY OF THE ARMY OF THE POTOMAC *by Charles D. Rhodes*—Including Pope's Army of Virginia and the Cavalry Operations in West Virginia During the American Civil War.

CAMP-FIRE AND COTTON-FIELD *by Thomas W. Knox*—A New York Herald Correspondent's View of the American Civil War.

SERGEANT STILLWELL *by Leander Stillwell* —The Experiences of a Union Army Soldier of the 61st Illinois Infantry During the American Civil War.

STONEWALL'S CANNONEER *by Edward A. Moore*—Experiences with the Rockbridge Artillery, Confederate Army of Northern Virginia, During the American Civil War.

THE SIXTH CORPS *by George Stevens*—The Army of the Potomac, Union Army, During the American Civil War.

THE RAILROAD RAIDERS *by William Pittenger*—An Ohio Volunteers Recollections of the Andrews Raid to Disrupt the Confederate Railroad in Georgia During the American Civil War.

CITIZEN SOLDIER *by John Beatty*—An Account of the American Civil War by a Union Infantry Officer of Ohio Volunteers Who Became a Brigadier General.

COX: PERSONAL RECOLLECTIONS OF THE CIVIL WAR--VOLUME 1 *by Jacob Dolson Cox*—West Virginia, Kanawha Valley, Gauley Bridge, Cotton Mountain, South Mountain, Antietam, the Morgan Raid & the East Tennessee Campaign.

COX: PERSONAL RECOLLECTIONS OF THE CIVIL WAR--VOLUME 2 *by Jacob Dolson Cox*—Siege of Knoxville, East Tennessee, Atlanta Campaign, the Nashville Campaign & the North Carolina Campaign.

KERSHAW'S BRIGADE VOLUME 1 *by D. Augustus Dickert*—Manassas, Seven Pines, Sharpsburg (Antietam), Fredricksburg, Chancellorsville, Gettysburg, Chickamauga, Chattanooga, Fort Sanders & Bean Station.

KERSHAW'S BRIGADE VOLUME 2 *by D. Augustus Dickert*—At the wilderness, Cold Harbour, Petersburg, The Shenandoah Valley and Cedar Creek..

AVAILABLE ONLINE AT **www.leonaur.com**
AND FROM ALL GOOD BOOK STORES

07/09

LEONAUR

ALSO FROM LEONAUR
AVAILABLE IN SOFTCOVER OR HARDCOVER WITH DUST JACKET

THE RELUCTANT REBEL by William G. Stevenson—A young Kentuckian's experiences in the Confederate Infantry & Cavalry during the American Civil War..

BOOTS AND SADDLES by Elizabeth B. Custer—The experiences of General Custer's Wife on the Western Plains.

FANNIE BEERS' CIVIL WAR by Fannie A. Beers—A Confederate Lady's Experiences of Nursing During the Campaigns & Battles of the American Civil War.

LADY SALE'S AFGHANISTAN by Florentia Sale—An Indomitable Victorian Lady's Account of the Retreat from Kabul During the First Afghan War.

THE TWO WARS OF MRS DUBERLY by Frances Isabella Duberly—An Intrepid Victorian Lady's Experience of the Crimea and Indian Mutiny.

THE REBELLIOUS DUCHESS by Paul F. S. Dermoncourt—The Adventures of the Duchess of Berri and Her Attempt to Overthrow French Monarchy.

LADIES OF WATERLOO by Charlotte A. Eaton, Magdalene de Lancey & Juana Smith—The Experiences of Three Women During the Campaign of 1815: Waterloo Days by Charlotte A. Eaton, A Week at Waterloo by Magdalene de Lancey & Juana's Story by Juana Smith.

TWO YEARS BEFORE THE MAST by Richard Henry Dana. Jr.—The account of one young man's experiences serving on board a sailing brig—the Penelope—bound for California, between the years 1834-36.

A SAILOR OF KING GEORGE by Frederick Hoffman—From Midshipman to Captain—Recollections of War at Sea in the Napoleonic Age 1793-1815.

LORDS OF THE SEA by A. T. Mahan—Great Captains of the Royal Navy During the Age of Sail.

COGGESHALL'S VOYAGES: VOLUME 1 by George Coggeshall—The Recollections of an American Schooner Captain.

COGGESHALL'S VOYAGES: VOLUME 2 by George Coggeshall—The Recollections of an American Schooner Captain.

TWILIGHT OF EMPIRE by Sir Thomas Ussher & Sir George Cockburn—Two accounts of Napoleon's Journeys in Exile to Elba and St. Helena: Narrative of Events by Sir Thomas Ussher & Napoleon's Last Voyage: Extract of a diary by Sir George Cockburn.

AVAILABLE ONLINE AT **www.leonaur.com**
AND FROM ALL GOOD BOOK STORES

07/09

LEONAUR
ALSO FROM LEONAUR
AVAILABLE IN SOFTCOVER OR HARDCOVER WITH DUST JACKET

ESCAPE FROM THE FRENCH *by Edward Boys*—A Young Royal Navy Midshipman's Adventures During the Napoleonic War.

THE VOYAGE OF H.M.S. PANDORA *by Edward Edwards R. N. & George Hamilton, edited by Basil Thomson*—In Pursuit of the Mutineers of the Bounty in the South Seas—1790-1791.

MEDUSA *by J. B. Henry Savigny and Alexander Correard and Charlotte-Adélaïde Dard* —Narrative of a Voyage to Senegal in 1816 & The Sufferings of the Picard Family After the Shipwreck of the Medusa.

THE SEA WAR OF 1812 VOLUME 1 *by A. T. Mahan*—A History of the Maritime Conflict.

THE SEA WAR OF 1812 VOLUME 2 *by A. T. Mahan*—A History of the Maritime Conflict.

WETHERELL OF H. M. S. HUSSAR *by John Wetherell*—The Recollections of an Ordinary Seaman of the Royal Navy During the Napoleonic Wars.

THE NAVAL BRIGADE IN NATAL *by C. R. N. Burne*—With the Guns of H. M. S. Terrible & H. M. S. Tartar during the Boer War 1899-1900.

THE VOYAGE OF H. M. S. BOUNTY *by William Bligh*—The True Story of an 18th Century Voyage of Exploration and Mutiny.

SHIPWRECK! *by William Gilly*—The Royal Navy's Disasters at Sea 1793-1849.

KING'S CUTTERS AND SMUGGLERS: 1700-1855 *by E. Keble Chatterton*—A unique period of maritime history-from the beginning of the eighteenth to the middle of the nineteenth century when British seamen risked all to smuggle valuable goods from wool to tea and spirits from and to the Continent.

CONFEDERATE BLOCKADE RUNNER *by John Wilkinson*—The Personal Recollections of an Officer of the Confederate Navy.

NAVAL BATTLES OF THE NAPOLEONIC WARS *by W. H. Fitchett*—Cape St. Vincent, the Nile, Cadiz, Copenhagen, Trafalgar & Others.

PRISONERS OF THE RED DESERT *by R. S. Gwatkin-Williams*—The Adventures of the Crew of the Tara During the First World War.

U-BOAT WAR 1914-1918 *by James B. Connolly/Karl von Schenk*—Two Contrasting Accounts from Both Sides of the Conflict at Sea D uring the Great War.

AVAILABLE ONLINE AT **www.leonaur.com**
AND FROM ALL GOOD BOOK STORES
07/09

LEONAUR

ALSO FROM LEONAUR
AVAILABLE IN SOFTCOVER OR HARDCOVER WITH DUST JACKET

IRON TIMES WITH THE GUARDS *by An O. E. (G. P. A. Fildes)*—The Experiences of an Officer of the Coldstream Guards on the Western Front During the First World War.

THE GREAT WAR IN THE MIDDLE EAST: 1 *by W. T. Massey*—The Desert Campaigns & How Jerusalem Was Won---two classic accounts in one volume.

THE GREAT WAR IN THE MIDDLE EAST: 2 *by W. T. Massey*—Allenby's Final Triumph.

SMITH-DORRIEN *by Horace Smith-Dorrien*—Isandlwhana to the Great War.

1914 *by Sir John French*—The Early Campaigns of the Great War by the British Commander.

GRENADIER *by E. R. M. Fryer*—The Recollections of an Officer of the Grenadier Guards throughout the Great War on the Western Front.

BATTLE, CAPTURE & ESCAPE *by George Pearson*—The Experiences of a Canadian Light Infantryman During the Great War.

DIGGERS AT WAR *by R. Hugh Knyvett & G. P. Cuttriss*—"Over There" With the Australians by R. Hugh Knyvett and Over the Top With the Third Australian Division by G. P. Cuttriss. Accounts of Australians During the Great War in the Middle East, at Gallipoli and on the Western Front.

HEAVY FIGHTING BEFORE US *by George Brenton Laurie*—The Letters of an Officer of the Royal Irish Rifles on the Western Front During the Great War.

THE CAMELIERS *by Oliver Hogue*—A Classic Account of the Australians of the Imperial Camel Corps During the First World War in the Middle East.

RED DUST *by Donald Black*—A Classic Account of Australian Light Horsemen in Palestine During the First World War.

THE LEAN, BROWN MEN *by Angus Buchanan*—Experiences in East Africa During the Great War with the 25th Royal Fusiliers—the Legion of Frontiersmen.

THE NIGERIAN REGIMENT IN EAST AFRICA *by W. D. Downes*—On Campaign During the Great War 1916-1918.

THE 'DIE-HARDS' IN SIBERIA *by John Ward*—With the Middlesex Regiment Against the Bolsheviks 1918-19.

AVAILABLE ONLINE AT **www.leonaur.com**
AND FROM ALL GOOD BOOK STORES

07/09

LEONAUR

ALSO FROM LEONAUR

AVAILABLE IN SOFTCOVER OR HARDCOVER WITH DUST JACKET

FARAWAY CAMPAIGN *by F. James*—Experiences of an Indian Army Cavalry Officer in Persia & Russia During the Great War.

REVOLT IN THE DESERT *by T. E. Lawrence*—An account of the experiences of one remarkable British officer's war from his own perspective.

MACHINE-GUN SQUADRON *by A. M. G.*—The 20th Machine Gunners from British Yeomanry Regiments in the Middle East Campaign of the First World War.

A GUNNER'S CRUSADE *by Antony Bluett*—The Campaign in the Desert, Palestine & Syria as Experienced by the Honourable Artillery Company During the Great War .

DESPATCH RIDER *by W. H. L. Watson*—The Experiences of a British Army Motorcycle Despatch Rider During the Opening Battles of the Great War in Europe.

TIGERS ALONG THE TIGRIS *by E. J. Thompson*—The Leicestershire Regiment in Mesopotamia During the First World War.

HEARTS & DRAGONS *by Charles R. M. F. Crutwell*—The 4th Royal Berkshire Regiment in France and Italy During the Great War, 1914-1918.

INFANTRY BRIGADE: 1914 *by John Ward*—The Diary of a Commander of the 15th Infantry Brigade, 5th Division, British Army, During the Retreat from Mons.

DOING OUR 'BIT' *by Ian Hay*—Two Classic Accounts of the Men of Kitchener's 'New Army' During the Great War including *The First 100,000 & All In It*.

AN EYE IN THE STORM *by Arthur Ruhl*—An American War Correspondent's Experiences of the First World War from the Western Front to Gallipoli-and Beyond.

STAND & FALL *by Joe Cassells*—With the Middlesex Regiment Against the Bolsheviks 1918-19.

RIFLEMAN MACGILL'S WAR *by Patrick MacGill*—A Soldier of the London Irish During the Great War in Europe including *The Amateur Army*, *The Red Horizon & The Great Push*.

WITH THE GUNS *by C. A. Rose & Hugh Dalton*—Two First Hand Accounts of British Gunners at War in Europe During World War 1- Three Years in France with the Guns and With the British Guns in Italy.

THE BUSH WAR DOCTOR *by Robert V. Dolbey*—The Experiences of a British Army Doctor During the East African Campaign of the First World War.

AVAILABLE ONLINE AT www.leonaur.com
AND FROM ALL GOOD BOOK STORES
07/09

LEONAUR

ALSO FROM LEONAUR

AVAILABLE IN SOFTCOVER OR HARDCOVER WITH DUST JACKET

THE 9TH—THE KING'S (LIVERPOOL REGIMENT) IN THE GREAT WAR 1914 - 1918 *by Enos H. G. Roberts*—Mersey to mud—war and Liverpool men.

THE GAMBARDIER *by Mark Severn*—The experiences of a battery of Heavy artillery on the Western Front during the First World War.

FROM MESSINES TO THIRD YPRES *by Thomas Floyd*—A personal account of the First World War on the Western front by a 2/5th Lancashire Fusilier.

THE IRISH GUARDS IN THE GREAT WAR - VOLUME 1 *by Rudyard Kipling*—Edited and Compiled from Their Diaries and Papers—The First Battalion.

THE IRISH GUARDS IN THE GREAT WAR - VOLUME 1 *by Rudyard Kipling*—Edited and Compiled from Their Diaries and Papers—The Second Battalion.

ARMOURED CARS IN EDEN *by K. Roosevelt*—An American President's son serving in Rolls Royce armoured cars with the British in Mesopatamia & with the American Artillery in France during the First World War.

CHASSEUR OF 1914 *by Marcel Dupont*—Experiences of the twilight of the French Light Cavalry by a young officer during the early battles of the great war in Europe.

TROOP HORSE & TRENCH *by R.A. Lloyd*—The experiences of a British Lifeguardsman of the household cavalry fighting on the western front during the First World War 1914-18.

THE EAST AFRICAN MOUNTED RIFLES *by C.J. Wilson*—Experiences of the campaign in the East African bush during the First World War.

THE LONG PATROL *by George Berrie*—A Novel of Light Horsemen from Gallipoli to the Palestine campaign of the First World War.

THE FIGHTING CAMELIERS *by Frank Reid*—The exploits of the Imperial Camel Corps in the desert and Palestine campaigns of the First World War.

STEEL CHARIOTS IN THE DESERT *by S. C. Rolls*—The first world war experiences of a Rolls Royce armoured car driver with the Duke of Westminster in Libya and in Arabia with T.E. Lawrence.

WITH THE IMPERIAL CAMEL CORPS IN THE GREAT WAR *by Geoffrey Inchbald*—The story of a serving officer with the British 2nd battalion against the Senussi and during the Palestine campaign.

AVAILABLE ONLINE AT www.leonaur.com
AND FROM ALL GOOD BOOK STORES

07/09

www.ingramcontent.com/pod-product-compliance
Lightning Source LLC
Chambersburg PA
CBHW031856090426
42741CB00005B/526